Sulfur Research Trends

The Third Annual Mardi Gras
Symposium Sponsored by the
Louisiana Section of the
American Chemical Society
at Loyola University, New
Orleans, La., Feb. 18–19, 1971.

David J. Miller

T. K. Wiewiorowski

Symposium Chairmen

ADVANCES IN CHEMISTRY SERIES **110**

AMERICAN CHEMICAL SOCIETY

WASHINGTON, D. C.   1972

ADCSAJ 110 1–231 (1972)

Library of Congress Catalog Card 72-77163

ISBN 8412-0137-4

PRINTED IN THE UNITED STATES OF AMERICA

Advances in Chemistry Series

Robert F. Gould, *Editor*

FOREWORD

ADVANCES IN CHEMISTRY SERIES was founded in 1949 by the American Chemical Society as an outlet for symposia and collections of data in special areas of topical interest that could not be accommodated in the Society's journals. It provides a medium for symposia that would otherwise be fragmented, their papers distributed among several journals or not published at all. Papers are refereed critically according to ACS editorial standards and receive the careful attention and processing characteristic of ACS publications. Papers published in ADVANCES IN CHEMISTRY SERIES are original contributions not published elsewhere in whole or major part and include reports of research as well as reviews since symposia may embrace both types of presentation.

CONTENTS

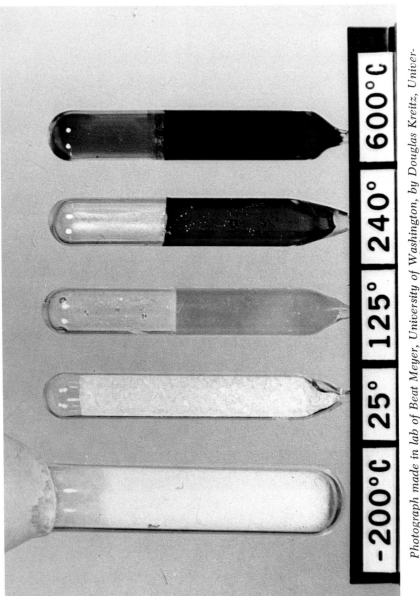

-200°C   25°   125°   240°   600°C

Photograph made in lab of Beat Meyer, University of Washington, by Douglas Kreitz, University of California at Berkeley.

PREFACE

In recent years, there has been a noticeable intensification in sulfur research. Current efforts cover a wide area of research, ranging from theoretical calculations on bonding and structure of sulfur and its compounds to studies on applied sulfur chemistry frequently oriented toward the development of new uses for this element.

The Third Annual Mardi Gras Symposium, organized by the Louisiana Section of the American Chemical Society, provided an opportunity to review the current trends in sulfur research. The previous two Mardi Gras Symposia were devoted entirely to theoretical chemistry. Sulfur chemistry became the major theme of this annual Symposium series for the first time this year since we felt that a need existed for a forum where sulfur researchers could report their latest findings and where they could informally exchange their knowledge concerning the various facets of sulfur chemistry—ranging from theoretical to the practical and applied aspects of the subject.

The Louisiana Section gratefully acknowledges the financial contributions made to the Symposium by Loyola University, Louisiana State University—New Orleans, Tulane University, and The Sulphur Institute.

The success of the Symposium must be largely attributed to R. L. Flurry, L. P. Gary, Jr., and O. E. Weigang, Jr., who served on the Program Committee and to S. P. McGlynn, B. Meyer, G. Griffin, and H. L. Fike who assumed the roles of discussion leaders during the Symposium.

A two-day Symposium certainly cannot give a full account of the status of sulfur chemistry. Yet, we believe that the papers presented in this volume are indicative of the current trends and of the progress being made in this field. Finally, we are convinced that the spirit of the Mardi Gras Season provided a stimulating background for this Symposium and an atmosphere which proved conducive to lively, informal, and effective exchange of thought.

D. J. Miller and
T. K. Wiewiorowski

New Orleans, Louisiana
February 1971

1

Semiempirical Molecular Orbital Calculations on Sulfur Containing Molecules and Their Extension to Heavy Elements

L. C. CUSACHS[a] and D. J. MILLER

Department of Computer Science, Loyola University, New Orleans, La. 70118 and Freeport Sulphur Co., Belle Chasse, La. 70118

Calculated or observed atomic orbital data, but no parameters derived from other calculations or measured properties of molecules, are used to construct a molecular orbital technique suitable for molecules containing unfamiliar elements. Predicted dipole moments for a series of Group IV/Group VI diatomic molecules and ionization potentials of SF₆ agree with the experimental values. A procedure analogous to the variation method in ab initio *calculations provides useful estimates of the 3d and 4s orbitals of sulfur and suggests suitable 4d orbitals for Se and 5d orbitals for Te. The dipole moment calculations provide evidence that relativistic energy contributions may be chemically significant, particularly for Te.*

The possibility of finding new electronic, physical, or chemical properties in molecules containing heavy elements is an incentive for exploring the region of the periodic table where relativistic effects become chemically significant. In a series of papers (*1, 2, 3, 4, 5, 6, 7, 8, 9, 10, 11, 12, 13, 14, 15, 16, 17, 18*) we have developed features of a semiempirical molecular orbital method that is as well defined for the heavy elements as for the familiar light ones. In this paper we review the theoretical foundation of this method, sources of atomic data necessary as input to the molecular calculation, and quality of results obtained at the present state of the art.

[a] Present address: Department of Computer Science, Loyola University, New Orleans, La. 70118.

The purpose of our semiempirical procedure is well described by Slater, Mann, Wilson, and Wood (19): we wish to calculate orbital energies for the occupied molecular orbitals which will be useful approximations to the corresponding ionization potentials of the molecules. For the molecular orbitals empty in the normal state of the molecule, we seek energies such that the difference between filled and empty orbital energies approximates the average of the corresponding singlet and triplet transition energies. We want the molecular orbitals to be useful for calculating dipole moments and oscillator strengths and in elucidating the chemistry of the molecule. Unavoidable many-electron effects, such as the splitting of singlet and triplet excited states, we relegate to secondary calculations which are not discussed further here. As noted by Slater et al. (19), theoreticians devoted to the Hartree–Fock approximation have often been unduly pessimistic about the possibility of success in this effort.

Our procedure assumes that we know how to obtain valence atomic orbitals and their ionization potentials and attempts to use this information to predict the corresponding molecular orbitals and their ionization potentials. Pseudopotential theory, developed originally for analogous calculations for solids, is readily adapted to our purpose and is a natural theoretical framework because the atomic data required are obtainable.

We refrain from trying to "improve" our calculations by incorporating parameters determined by forcing agreement with experiment for selected properties of small molecules for two reasons:

(1) we want our results to be genuine predictions, rather than interpolations or extrapolations,

(2) extensive data for even small molecules containing many of the heavier elements are frequently unavailable.

Atomic data are used extensively in this approximate method. We consider the three basic sources of data defining the atomic orbitals required; they are experimental valence state ionization potentials, accurate atomic calculations, and determination in the molecule by a process analogous to variation in ab initio calculations. The determination of extravalent orbitals for sulfur, selenium, and tellurium is described as an application of the last procedure.

The electric dipole moment of a molecule is a sensitive property that, in principle, may be calculated accurately in a one-electron model. Since good results for any single molecule might result easily from accidental cancellation of errors, we include results for all of the Group IV/Group VI diatomics for which we have found recent reliable experimental values for comparison. Our dipole moments are obtained by calculating the expectation value of the operator without resort to point–charge or similar approximations. As expected from the band theory literature, we find that orbital energy data from nonrelativistic atomic calculations must be corrected for relativistic terms for the heavier elements, including Pb and Te, and probably also for Sn and Se.

We have accurate observed values for only a few molecules for the ionization potentials corresponding to all the molecular orbitals built up from valence shell atomic orbitals. SF_6 is interesting since probably the complete set of valence orbital ionization potentials have been measured and identified. Our calculations agree well with the experiment. We consider also CO_2 and SO_2 where the data available to us are not complete and the agreement between our calculations and reported observations is not so completely in accord.

Theoretical Foundation

Starting with a set of atomic orbitals, ϕ_i, we generate molecular orbitals (1) as linear combinations of the atomic orbitals, $\Phi_m = \Sigma_i C_{mi}\phi_i$, by seeking the coefficients, C_{mi}, which minimize the total energy. This leads to the usual secular equation, in matrix form, $HC = SCE$. This determination requires two types of matrix element over the atomic orbitals, overlap integrals, $S_{ij} = <i|j>$, and energy integrals, $H_{ij} = <i|H|j>$. Since we are not simply attempting a shortcut to reduce the labor of a minimum basis set *ab initio* calculation, the atomic orbitals will, in a special sense, represent the most accurate atomic orbitals we can find. There is little apparent problem with overlap integrals since they are calculated directly using a single Slater-Type orbital, or STO, chosen (3) to reproduce best overlap integrals that would be obtained with accurate atomic orbitals. The H_{ij} are not obtained by more devious means because we cannot expect to find a set of single STO per atomic orbitals from which both overlap and energy integrals could be calculated. The reason is simple (4, 5); to reproduce in the best way overlap integrals that would be obtained from accurate solutions of the atomic Hartree–Fock equations, for instance, the single STO must have the same moments, $<r>$ and $<r^2>$. To reproduce energy integrals, it must reproduce $<1/r>$ and $<1/r^2>$ as well, requirements that are generally incompatible.

Table I gives the atomic orbital data used in our calculations (4, 5). Table II gives values of the effective principal quantum numbers of a selection of valence atomic orbitals obtained by requiring that the product of $<r>$ and $<1/r>$ match that of the Hartree–Fock functions (20, 22). These are the basis orbitals, ϕ_i, used to expand the valence molecular orbitals, Φ_m. The standard STO has a radial part proportional to

$$r^{n-1} \exp(-zr)$$

containing two parameters, the principal quantum number, n, and the orbital exponent or *zed*, z. For such an orbital the average radius, $<r>$,

Table I. Atomic Orbital Data

Atom	Orbital	n	STO z	A_i CFF[b]	A_i REL[c]	A_i VSIP[d]	bohr[e] RI
C	2s	2	1.57	19.37	—	19.54	1.12
C	2p	1	0.88	11.07	—	9.69	1.28
Si	3s	3	1.58	14.79	—	14.76	1.65
Si	3p	2	0.91	7.58	—	7.90	2.22
Ge	4s	3	1.58	15.15	—	—	1.72
Ge	4p	2	0.87	7.33	—	—	2.26
Sn	5s	3	1.36	13.04	—	—	2.04
Sn	5p	3	1.06	6.76	—	—	2.61
Pb	6s	3	1.29	12.48	16.66	—	2.17
Pb	6p	3	1.01	6.52	7.64	—	2.72
O	2s	2	2.19	34.01	—	32.00	0.79
O	2p	1	1.22	16.76	—	15.30	0.90
S	3s	3	2.03	24.01	—	21.29	1.28
S	3p	2	1.21	11.60	—	11.38	1.54
S[a]	3s	2	1.53	—	—	21.29	1.28
S[a]	3p	2	1.21	—	—	12.00	1.54
S[a]	3d	3	1.70	—	—	2.00	—
Se	4s	3	1.87	22.85	23.93	—	1.44
Se	4p	2	1.09	10.68	11.00	—	1.78
Se[a]	4s	3	1.87	22.77	—	—	1.44
Se[a]	4p	2	1.09	10.95	—	—	1.78
Se[a]	4d	4	a	0.0	—	—	—
Te	5s	3	1.57	19.11	21.41	—	1.74
Te	5p	3	1.30	9.54	10.22	—	2.12
Te[a]	5s	3	1.57	19.14	—	—	1.74
Te[a]	5p	3	1.30	9.54	—	—	2.12
Te[a]	5d	5	a	0.0	—	—	—
B	2s	2	1.26	13.46	—	—	1.40
B	2p	1	0.68	8.43	—	—	1.65
Al	3s	2	0.96	10.70	—	11.32	1.97
Al	3p	2	0.73	5.71	—	5.98	2.64
Ga	4s	2	1.41	11.55	—	—	1.94
Ga	4p	2	0.73	5.67	—	—	2.69
In	5s	3	1.67	10.14	—	—	1.64
In	5p	3	1.40	5.37	—	—	1.96
F	2s	2	2.50	42.77	—	39.00	0.70
F	2p	1	1.38	19.86	—	18.20	0.79

[a] Used for X_4 and GIVNAP SF_6 calculations only.
[b] Data from C. Froese (20).
[c] Data from C. Froese (20) with relativistic corrections (21).
[d] Data from L. C. Cusachs and J. H. Corrington (4, 5).
[e] Data from C. Froese (20), available from J. B. Mann (22).

is $(n + 1/2)/z$ while $<1/r>$ is z/n. The product of these two average values, for a STO, is $(1 + 1/2n)$, independent of z. The product of these two moments for the Hartree–Fock orbitals thus defines the principal quantum number of a single STO that could reproduce both moments. The data of Table II are obtained by this calculation.

The effective principal quantum number in the regions that make major contributions to these integrals is generally less than the nominal principal quantum number of the orbital, a phenomenon encountered (*4, 5*) in determining single STO to match overlap integrals (Table II). It is known well in *ab initio* calculations that the use of two or more STO for each nominal atomic orbital produces dramatic improvement in molecular properties over a single function. This is not surprising when we realize that the minimum principal quantum number usable in *ab initio* calculation is limited by continuity conditions at the nucleus to $n = l + 1$, where l is the angular quantum number. From the atomic calculations available, it is not clear whether atomic orbitals outside of the valence shell should be represented by STO of the nominal or reduced principal quantum numbers.

Since we will be using energy integrals from atomic calculations or experiment rather than attempting to compute them, we take the oppor-

Table II.   Effective Principal Quantum Numbers

Atom	Orbital	Neff[a]	Orbital	Neff
Be	2s	1.30		
B	2s	1.22	2p	1.50
C	2s	1.18	2p	1.43
N	2s	1.14	2p	1.38
O	2s	1.12	2p	1.34
F	2s	1.11	2p	1.32
Si	3s	1.51	3p	1.56
S	3s	1.42	3p	1.46
Cl	3s	1.38	3p	1.43
Ge	4s	1.73	4p	1.77
Se	4s	1.67	4p	1.69
Br	4s	1.64	4p	1.65
Sn	5s	1.86	5p	1.92
Te	5s	1.81	5p	1.84
I	5s	1.79	5p	1.81
Pb	6s	1.98	6p	2.04
Po	6s	1.94	6p	1.97
Ni	3d	1.49	4s	1.80
Pt	5d	1.60	6s	2.02

[a] Neff is calculated from $<r>$ and $<1/r>$ for the orbital, using the tabulated moments for accurate Hartree–Fock atomic orbitals (*22*). These should not be confused with the effective values of the principal quantum number for overlap, which are determined from $<r>$ and $<r^2>$.

tunity to correct for some of the many-electron effects missing in the molecular orbital approximation by preferring valence state atomic data where available.

We leave the construction of the H matrix to consider core-valence separation. The treatment of the separation of the inner shells or cores from the valence orbitals is important in understanding what quantities we are trying to approximate with energy integrals deduced from atomic data.

There is much precedent but no particular justification for omitting the core orbitals from the molecular calculation. To determine the consequences of (6) separating the core orbitals and electrons, we divide the set of atomic orbitals into two classes, X_p, core orbitals with energies E_p, and valence orbitals for which we retain the symbol ϕ_i. Unlike the C_{mi}, the coefficients of the core orbitals are not free for variation to minimize the energy but are determined by requiring that arbitrary admixture of the core orbitals in the valence molecular orbitals do not change the energy of the latter. The final matrix equation (6) is of the order of the number of valence orbitals, but the definitions of the S and H matrix elements are modified:

$$H_{ij} = \overset{\circ}{H}_{ij} - \Sigma_p E_p S_{ip} S_{jp}$$
$$S_{ij} = \overset{\circ}{S}_{ij} - \Sigma_p S_{ip} S_{jp}$$

The advantage of this procedure is that the valence orbitals need not be orthogonal to the core orbitals—*i.e.*, we can use nodeless functions for the valence orbitals. The price is that the elements of the H and S matrices are not what we should calculate directly without regard to the core orbitals but the expressions above for H and S. This is an advantage rather than a difficulty. To show this, the normalization of the atomic orbitals is chosen such that S_{ii} (rather than $\overset{\circ}{S}_{ii}$) $= 1$. Tables have been presented (1) showing that even with a more severe correction, typical two-center overlap integrals are not modified seriously.

The condition leading to the above equations is that arbitrary admixture of core orbitals does not affect the valence orbital energies. Therefore the H_{ii} rather than the $\overset{\circ}{H}_{ii}$ are identified with the energy integrals over the accurate atomic orbitals. The difference between using semi-empirical values of H_{ii} and calculating integrals over valence orbitals orthogonalized to all cores on the same atom is that we replace the kinetic energy contribution from the nodes by an effective potential. Since only the sum is of interest rather than individual kinetic and potential terms, there is no need to separate them. However, if we compare individual integrals with *ab initio* calculations, we would have to find some way to determine $\overset{\circ}{H}_{ij}$. We believe this is the reason that attempts to understand the Extended Hückel and other methods using single STO for overlap, and H_{ii} determined from empirical atomic data by comparing individual terms with Hartree–Fock matrix elements have been less successful than the semiempirical methods themselves.

The above equations, which we call the LCAO pseudo-potential equations, also tell us how to include extravalent orbitals (7) such as the 4s orbital of sulfur in the same calculation as the valence orbitals such as the 3s of sulfur.

This core separation was called pseudopotential theory since the effective potential contains kinetic as well as potential energy contributions (23).

Within our one-electron model, electron repulsion enters as one of several contributions to the effective potential. The diagonal elements of H, the H_{ii}, are assumed to be simple linear functions of atom net charge plus the neighbor atom potential (8, 9, 10) which is representing the potential resulting from the distribution of positive and negative charge on the other atoms in the molecule. The neighbor atom potential contains nuclear attraction and electron repulsion integrals, both sufficiently complicated owing to precautions for maintaining rotational invariance that we refer the reader to reference papers for details (1, 2, 8, 9, 10). The integrals in the neighbor atom potential require an effective radius, $RI = 1/<1/r>$, for each atomic orbital which may be used also to estimate the variation of the H_{ii} with atom charge, meaning that only two energy quantities or parameters are required.

We define H_{ii} as the negative of the ionization potential of the atomic orbital, doubly occupied, in the molecular environment.

Sources of atomic data for determining the H_{ii} are discussed at length below. There are two contributions to the $H_{ij}(i \neq j)$. The neighbor atom potential introduces terms connecting members of a set of p, d, or f orbitals on the same atom if the neighbor atom is not in the direction of a coordinate axis from the reference atom. These H_{ij} terms are necessary to preserve rotational invariance. For atomic orbitals on different atoms,

$$H_{ij} = 0.5(\hat{H}_{ii} + \hat{H}_{jj})\,(2S_{ij}|\hat{S}_{ij}|)$$

where the special symbols H and S mean:

(1) To preserve rotational invariance, the H_{ii} used to estimate H_{ij} must be averaged over orientations for p, d, or f orbitals, and only for this purpose.

(2) the quantity $S_{ij}|S_{ij}|$ is calculated in a local diatomic coordinate system and related to the master coordinate system of the molecule treating it as transforming under rotation like an ordinary overlap integral.

Our method is compared profitably with the extended Hückel method of Roald Hoffmann (24, 25). We differ in two major respects. We consider the effects of charge redistribution, which is important in polar molecules, forcing us to seek charge self-consistency iteratively. We attempt to find atomic orbitals that represent quantitatively the relevant properties of more complicated and more accurate atomic orbitals. To describe the charge redistribution and neighbor atom potential, we require one additional item of atomic orbital data. The computation is an order of magnitude greater, for which we obtain:

(1) in the cases so far studied, quantitatively better ionization potentials and dipole moments.

(2) approximate electron repulsion integrals which permit us to construct a valence total energy and to attack electronic spectra.

The difficulty of a particular computer calculation is judged best by the number of instructions in the program or the computer time required for its execution. The handling of the H_{ij} expression above is easy while the computation of electron interaction integrals, simpler by inspection, is more difficult.

Hoffmann's extended Hückel method (24, 25) is regarded as a first approximation to our method, assuming a constant effective potential and within the limitations of his choice of atomic orbital data and H_{ij} approximation. It is valid where charge redistribution is neglected. A second approximation within the same theoretical framework is to adjust the H_{ii} for gain or loss of charge but not explicitly introducing electron interaction integrals. The calculations of Carroll, Armstrong, and McGlynn (26) for H_2O and H_2S adopted this procedure using atomic valence state data as input. The iterative extended Hückel method of Rein et al. (27) attempts to include the average effects of neighbor atom potentials by calibrating the parameters A and B (below) to reproduce experimental data for model molecules. More sophisticated methods which also use calibration to compensate to a certain extent for the neglect of overlap integrals in the solution of the secular equation and other deficiencies are modeled on the CNDO method of Pople, Santry, and Segal (28). Many of these approximate Hartree–Fock methods have been reviewed by Dahl and Ballhausen (29). However, we do not share any of their enthusiasm for the CNDO approximation or for accepting the Hartree–Fock approximation as a formal theoretical foundation for semiempirical procedures. We believe that reliability in most chemical applications of molecular orbital theory requires a realistic treatment of the one-electron terms in the H matrix and the retention of overlap integrals in the secular equation. We do not see that the details of the treatment of electron interaction are important, except possibly for electronic spectra calculations. Methods requiring calibration are expected to produce relatively good results for the types of molecules and specific properties for which they are calibrated. When either the type of molecule or the property computed is changed significantly, the results are often disappointing.

Atomic Orbital Data

We assume for the diagonal matrix elements, H_{ii}, the form $H_{ii} = -A_i - B_i q^{net} + NAP$, where A_i is the ionization potential of the orbital doubly occupied in the free atom, B_i is the change of ionization potential with unit change in the net charge on the atom at which ϕ_i is centered, and NAP is the effective potential resulting from the other atoms in the molecule or neighbor atom potential (8, 9, 10). For many of the elements in the first two full rows of the periodic table suitable values of A have been determined (11, 12, 13) from atomic energy level data. For hydrogen we lack an experimental or calculated value for the ionization potential of the orbital, which is occupied doubly in the neutral free atom. For the lower half of the periodic table our atomic energy level data are much less complete, and we must resort to other sources for values of A and B. One source is the numerical Hartree–Fock atomic orbitals of Froese (20) or those of Mann (22). However these calculations were performed for average states of configurations rather than for states of definite orbital occupation or valence states; therefore, the numerical values differ from the empirical VSIP in some cases by as much as 1 ev. Values for some elements for which data from both sources are available appear in Table I. Fortunately for calculations for heavier elements, the distinction between single and double occupation of a particular orbital decreases in a family with increasing atomic weight. A much more important consideration for the transition elements is the configuration—*i.e.*, the distribution of electrons between different valence subshells. For instance the calculated and observed ionization potentials for the $3d$ orbital of Ni in the configurations $3d^8 4s^2$ and $3d^{10}$ differ by at least 12 ev. Limited experience in molecular calculations suggests that the $3d^{10}$ configuration is the appropriate one, but it is unfortunately not the one for which atomic calculations are available except for Ni since the tradition is to perform the calculations for the standard configuration $nd^{m-2}(n+1)s^2$ rather than nd^m for m electron transition elements. For elements of the sulfur and carbon families the orbital energies appear to be insensitive to the choice of configuration.

The quantity B_i also introduces problems. For the lighter elements both calculated and observed ionization potentials or orbital energies are available for positive ions as well as the neutral atoms. For the heavier elements the value of B is deduced from other information. B is regarded (11, 12, 13) as an effective value of an electron repulsion integral corrected for kinetic energy changes and orbital rearrangement. In developing a useful neighbor atom potential Corrington was forced to estimate nuclear attraction and electron repulsion integrals (5, 9), and her formula, in principle, is applicable to integrals between orbitals on the same atom

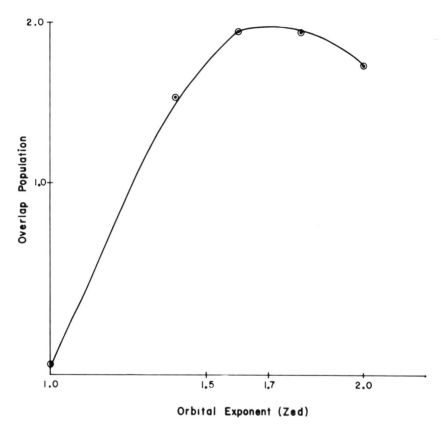

Figure 1.   Overlap population of S_4 as a function of zed 3d

as well as to orbitals on different atoms. Her integrals depend on an
effective orbital radius, $RI = 1/<1/r>$ for the orbital. If we suppose
the neighbor atom potential integrals should apply also to the same atom,
we can relate B to RI, reducing the parameters necessary for the energy
integrals to A and RI for each orbital. The RI are generally available
(4, 5, 20, 22), except for the desired configurations of d series transition
metals from the accurate atom HF calculations. This procedure has been
adopted in the calculations reported below.

The Extravalent d Orbitals of S, Se, and Te

Since the extravalent d orbitals of sulfur, selenium, and tellurium
are not occupied in the normal states of the free atom, their implicit
functional form is determined by another criterion. The values of a number of molecular properties including ionization potentials, energies of
electronic transitions, and overlap populations have been examined as a

function of d orbital exponent, *zed*, for a series of molecules with the general formula X_4 where $X = $ S, Se, and Te. We assume the molecules to be square with bond lengths of 2.04, 2.34, and 2.86 A, respectively.

Figure 1 gives a plot of the total overlap population obtained for the molecular S_4 as a function of *zed* $3d$. The TOP is maximum, and computed molecular properties are insensitive to the precise value of the orbital exponent, over a moderate range centered at 1.7. The value of *zed* $ = 1.7$ corresponds to a sulfur $3d$ orbital radius of 1.089 A, essentially the same radius as the $3p$ orbital.

Figure 2 is a plot of the total overlap population obtained for the molecular Se_4 as a function of zed_{4d}. If Kaufman's conjecture (*30*) concerning the maximization of total overlap population is accepted as a necessary condition to be met by our calculations, a value of $zed_{4d} = 2.2$ is predicted for selenium. Contrary to what is found in S_4, the curve at its maximum is not very flat, and the region over which computed molecular properties are insensitive to the precise value of zed_{4d} is very narrow. The d orbital radius corresponding to $zed_{4d} = 2.2$ is 2.28 A.

Preliminary studies for Te_4 indicate different orbital exponents, 2.12 or 2.45, depending on the choice of total overlap energy or total overlap population as the criterion of stationary property. Further calculations for Se_4 with a $3d$ STO produce a flatter curve similar to that of Figure 1.

Further computation should show that reduced principal quantum numbers should be used for all of these d orbitals. In the case of rhodium, for which no satisfactory orbitals are available to us from atomic calcula-

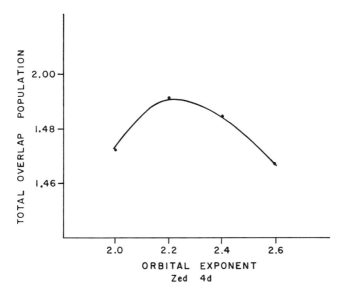

Figure 2.   Overlap population of Se_4 as a function of zed 4d

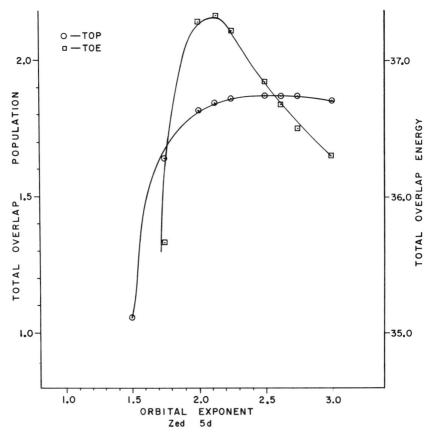

*Figure 3. Overlap population and overlap energy as functions of zed 5d
for Te₄*

tions, exploratory calculations for the species Rh_2, as yet unreported
experimentally, suggested (*16*) orbital exponents for the valence d and
outer s and p orbitals. The d and s functions were close to what was
expected from the previous results for S_4 (*15*) and H_2S (*14*), but the $5p$
function behaved differently when represented by $3p$ and $5p$ STO's. For
this reason calculations on Te_4 will be performed with the alternate
orbitals as computer time becomes available.

Dipole Moments

The recent appearance (*31*) of accurate experimental dipole moments
for a representative collection of Group IV/Group VI diatomic molecules
allows us to test the capabilities of our procedure. Since dipole moments
are predicted erratically by molecular orbital calculations, good agreement with one or two measured values would have little significance.

The general close agreement found in Table III suggests that the method reproduces the essential trends found by the experiments. At present we have a computer program capable of computing dipole moments from the wavefunctions for linear molecules, but these results are sufficiently encouraging that we plan to attempt the more general shapes of molecules as resources become available.

For the lighter elements we have available the VSIP data defined for orbitals of definite occupancy (*11, 12, 13*). For the heavier elements we have only the Hartree–Fock orbital energies (*20, 22*). Thus, for the lighter combinations we used either source of data while if either element is heavy, our choice is restricted. It seems preferable to use data

Table III.   Dipole Moments (Debyes)

Molecule	REL[a]	VSIP[b]	CFF[c]	EXP.[d]
CO		0.04	−0.00	−0.112
CS		−1.65	−2.35	−1.1985
CSe			−2.81	(−2.6)
CTe			−3.28	(−3.5)
SiO		2.79	3.22	3.098
SiS		1.29	1.48	1.73
SiSe	0.66		0.39	1.1
SiTe	−0.24		−0.67	(0.6)
GeO			4.10	3.282
GeS			2.29	2.00
GeSe	1.61		1.27	1.648
GeTe	0.92		0.09	1.06
SnO			6.05	4.32
SnS			3.65	3.18
SnSe	3.04		2.70	2.82
SnTe	2.55		1.56	2.19
PbO			7.55	4.64
PbS	2.89		4.46	3.59
PbSe	2.14		3.63	3.29
PbTe	1.31		2.39	2.73
BF			−0.95	−1.0
AlF		0.59	0.83	1.53
GaF			2.31	2.45
InF			3.16	3.41

[a] Relativistic contributions to the orbital energies computed by Herman and Skillman added to the nonrelativistic SCF orbital energies of Froese for use as VSIP.
[b] Valence state ionization energies used as neutral atom orbital ionization potentials.
[c] Hartree–Fock orbital energies of Froese used as neutral atom orbital ionization potentials.
[d] Experimental results from J. Hoeft *et al.* (*31*). Values in parentheses are estimated.
The calculated dipole moments are simply the expectation values of the dipole operator with respect to the wavefunction. The only approximation is the assumption that the inner shells exactly cancel a corresponding amount of nuclear charge.

from a single source since the variations tend to cancel partially. The relatively poor results for Pb may result from relativistic effects being so large that correction by perturbation theory is inadequate. A test of this conjecture waits on the availability of orbital energies and radial moments from the relativistic Hartree–Fock–Dirac model.

Beyond the limitations of the method itself and the uncertainties of the atomic data, there is an uncertainty resulting from the convergence tolerance in the charge self-consistency calculation that in some cases appears to be as large as 0.3 Debye or larger, particularly in difficult convergence cases, such as PbO, where we have had to accept a larger than usual difference between input and output of the final cycle. The usual tolerance is 0.20 ev in the H_{ii}, but we have had to raise it to as high as 0.5 in a few cases. We regard this limit as provisional for small molecules since it is possible that the extrapolation and damping process in the iterative scheme can be improved.

Ionization Potentials

The emergence of photoelectron spectroscopy (32, 33) offers prospects of accurate measurements of the ionization potentials for the entire set of valence molecular orbitals for many molecules. The measurement is direct, permitting resolution sufficient in many cases to assign the symmetry and predominant atomic orbital composition of the molecular orbital of the departing electron. Two kinds of ionization are reported— adiabatic and vertical. An adiabatic ionization removes an electron from an orbital of the molecule in its ground vibrational state and leaves it in the ground vibrational state of the positive ion. This is the type of ionization potential obtained by extrapolating to the limiting energy of Rydberg states of the molecule. The vertical ionization corresponds to removing the electron with the nuclei fixed in position for the ground vibrational state of the neutral molecule and leaving the ion in an excited vibrational state if the equilibrium geometry is different. In favorable cases either process is observed with great confidence. In interpreting complex photoelectron spectra, the adiabatic ionization potential is the energy of the first appreciable intensity at the low energy end of the spectral band while the vertical process should correspond to the maximum intensity. The difference between adiabatic and vertical ionization potentials is important for orbitals that are strongly bonding or strongly antibonding. Only the vertical ionization potentials represent processes for which we can calculate meaningful approximations.

When two or more ionizations occur close together, it is often difficult to be sure of the beginning or the maximum of a single ionization.

The case of SF_6 is useful for a comparison of molecular orbital calculations with experiment. The high symmetry of the molecule reduces

Table IV.   Orbital Energies and Ionization Potentials of SF_6

GIVNAP[a]			ARCANA		ESCA[b]	CNDO[c]	
3s/3p	3s/3p/3d		3s/3p	3s/3p/3d	exp.		
16.33	17.00	t_{1g}	16.23	16.63	16.	e_g	17.81
17.51	18.12	t_{1u}	17.42	17.79	17.3	t_{1u}	21.47
17.73	18.44	t_{2u}	17.62	18.06	17.3	t_{1g}	22.40
18.08	18.87	e_g	18.02	18.30	18.7	t_{2u}	23.14
20.21	21.01	t_{2g}	20.09	20.57	19.9	t_{2g}	24.23
22.66	22.60	t_{1u}	22.41	22.45	22.9	t_{1u}	29.01
25.03	24.88	a_{1g}	25.16	25.12	27.0	a_{1g}	30.80
37.94	38.78	e_g	37.78	38.36	39.3	e_g	43.14
40.28	40.53	t_{1u}	39.97	40.23	41.2	t_{1u}	50.79
46.89	45.92	a_{1g}	47.39	46.77	44.2	a_{1g}	57.03

[a] Calculations using the computer program GIVNAP, a preliminary version of ARCANA. Results for SF_6 from D. J. Miller (*34*). Effects of 3d orbitals may be overestimated. Input atomic data for the GIVNAP and ARCANA programs was identical; they differ principally in methods of generating nuclear attraction and electron repulsion integrals.

[b] Values for K. Seigbahn, *et al.* (*32*). The authors explain that the first two values reported include three ionizations, with the two probably near 17.3. They also cite unpublished *ab initio* calculations with d orbitals as giving the same level ordering as GIVNAP.

[c] Calculations without 3d orbitals for sulfur from b; b also states that CNDO calculations with 3d orbitals give an incorrect ordering.

the number of distinct valence molecular orbital ionization potentials to 10, and the observations and analysis afford a complete assignment of the ionization potentials. In Table IV we compare the results of calculations with ARCANA, its predecessor computer program GIVNAP, and a competitive many-parameter semiempirical method, CNDO, with the observed ionization potentials.

The agreement between our calculations and the observed ionization potentials is less complete in the other cases we have examined. In Table V we exhibit ARCANA results for SO_2 compared with accurate calculations that are possibly close to the present practical limit for *ab initio* calculations for a molecule this large and complex. The first observed ionization at 12.5 ev vertical, 12.29 adiabatic, is believed to be of symmetry type a_1. The next band is complicated by the superposition of two different ionizations, beginning at 12.98 ev, with maxima we judge to be at about 13.2 and 13.5 in the published spectra. Turner *et al.* (*33*) believe that this region contains ionizations from orbitals of a_2 and b_2 symmetry in general agreement with both calculation methods. The next band is difficult also, containing at least two ionizations of type a_1 and b_1. From the close agreement between the two sets of calculations we suspect that a third ionization potential, b_2, is present also.

Improvement in the calculation of ionization potentials is indicated by Table VI where we compare results for CO_2, COS, and CS_2 with other semiempirical calculations and the results of Turner *et al.* (*33*), this time with a problem in the ordering of the highest sigma and lowest pi symmetry levels. Since the highest pi and highest sigma levels are nonbonding, the vertical and adiabatic ionization potentials are essentially identical. The lower pi level is strongly bonding so its photoelectron spectrum is broad with an appreciable difference between the two types of process. The presence of the superimposed sigma ionization makes it difficult to be sure of the maximum. There appears to be a discrepancy in the ordering of levels as well as a systematic error in the actual values since the pi ionization potentials are too high and the sigma ones too low. To a lesser extent the same tendency is evident in Table V, though not sufficient to prejudice the calculation seriously.

Systematic errors in a semiempirical calculation of the type we have performed can be corrected by altering the input data. It is possible, for example, to alter the level ordering to that observed for CO_2 by tampering with the relation supposed between the B and RI parameters. It probably is possible also to effect the same sort of change by modifying the atomic orbitals used to calculate overlap integrals. Refinements, which might constitute improvements, are imagined easily for the H_{ij} approximation or in the inclusion of exchange integrals omitted presently from the calculations. We prefer, however, not to treat the atomic data as adjustable parameters but to explore the possibility of developing this calculation method as a bonafide prediction of molecular properties from those of the constituent atoms.

Table V.   Orbital Energies for Sulfur Dioxide

ARCANA			*EXP.*[a]	*HF–SCF*[b]		
s/p	*Sym.*	s/p/d		s/p	*Sym.*	s/p/d
11.12	a_1	11.53	12.5	13.41	a_1	13.37
13.10	b_2	13.80 ⎱	13.2, 13.5	13.79 ⎰	a_2	14.04
13.59	a_2	14.30 ⎰		15.08 ⎱	b_2	14.69
16.27	b_1	16.76 ⎤		18.95 ⎡	b_1	18.13
16.69	a_1	17.02 ⎬	16.6, 16.6	19.59 ⎨	a_1	18.93
17.71	b_1	17.71 ⎦		18.99 ⎣	b_2	19.00
21.13	a_1	21.34		24.25	a_1	23.65
29.96	b_2	30.88		39.36	b_2	38.18
34.55	a_1	35.00		42.80	a_1	41.06

[a] The experimental values are subject to considerable uncertainty because the bands overlap extensively. We have estimated several ionizations from the spectrum and discussions of D. W. Turner *et al.* (*33*).

[b] Gaussian calculations of S. Rothenberg and H. F. Schaefer III (*35*). The s/p orbital energies were furnished by Prof. Schaefer. The s/p/d set includes d orbitals on the oxygen atoms as well as the sulfur, while the ARCANA calculation s/p/d includes them only for sulfur.

Table VI.   Lower Ionization Potentials of CO_2, COS, and CS_2

	EXP^a	EXH^b	$CNDO^b$	$ARCANA$	
CO_2	13.78	17.20	14.55	14.90	πg
	17.59	18.14	16.83	18.09	πu
	18.08	17.51	14.55	17.35	σu
	19.40	19.56	21.01	20.44	σg
COS	11.18	13.14	12.26	12.75	π
	15.52	17.76	15.36	16.40	π
	16.04	14.67	13.81	14.92	σ
	17.96	18.40	19.32	17.97	σ
CS_2	10.08	12.24	11.33	11.28	πg
	12.83	14.18	13.28	14.86	πu
	14.47	13.48	12.59	13.37	σu
	16.19	15.72	18.23	15.65	σg

[a] Values from Turner, *et al.* (*33*). The authors interpret the spectra as far as symmetry assignments and values for adiabatic ionization potentials. We have attempted to estimate the vertical ionization potential for the second π ionization, which is 0.1–0.3 ev greater than the adiabatic since the vertical ionization is the process corresponding to the calculations.
[b] Data from J. M. Sichel, M. A. Whitehead (*36*).

The Sulfur–Oxygen Bond

The report by Van Wazer in this volume (Chapter 2) of calculations for the molecules H_2S, H_2SO, and H_2SO_2 prompts us to present our results for comparison. We considered first this series as part of a study of bonding in sulfoxides and sulfones, using an early version of the GIVNAP program. We have repeated them with ARCANA using the atom positions chosen by Van Wazer. The results are displayed in Table VII.

For hydrogen, which does not have the requisite valence state for use in our method, we have arbitrarily chosen a STO for overlap of principal quantum number 1 and *zed* = 1.2, and for energy terms A = 10.0 and *RI* = 1.2; while these values seem unlikely, they give reason-

Table VII.   Sulfur Hydrogen Bonds

Molecule	*Orbitals*	$q_H{}^L$	$q_H{}^M$	OP_{S-H}	OE_{S-H}
H_2S	s/p	+0.046	+0.067	0.749	−15.24
	$s/p/d$	+0.065	+0.036	0.722	−15.63
H_2SO	s/p	−0.004	+0.054	0.646	−15.20
	$s/p/d$	+0.034	+0.025	0.692	−15.49
H_2SO_2	s/p	−0.069	+0.009	0.576	−15.44
	$s/p/d$	−0.003	+0.003	0.687	−15.96

$q_H{}^L$ is the net charge on a hydrogen atoms in the Löwdin basis.
$q_H{}^M$ is the net charge on the hydrogen atoms by Mulliken population analysis.
OP_{S-H} is the overlap population between hydrogen and sulfur.
OE_{S-H} is the overlap energy between hydrogen and sulfur (ev).

able ionization potentials for the molecules containing hydrogen we have studied.

The startling feature of our and of Van Wazer's calculations is the degree to which the S–H bond is independent of the oxygen atoms. Calculations (37) for AgI complexes with coordinated NH_3 indicate that the positive charge assumed by the ammine ligand is borne mostly by the hydrogens rather than the nitrogen, exactly the reverse of the phenomenon encountered in these sulfur systems. We suggest that this difference between nitrogen and sulfur may be quite significant chemically.

Acknowledgments

We have drawn upon the efforts of our collaborators, Joyce H. Corrington of Xavier University of New Orleans, L. P. Gary of Loyola University, Willis Thornsberry of Freeport Sulphur Company, Haven S. Aldrich, and Clyde W. McCurdy of Tulane University. Part of this work was supported by a grant from the Sulphur Institute to Tulane University.

Literature Cited

(1) Cusachs, L. C., Cusachs, B. B., *J. Phys. Chem.* (1967) **71**, 1060.
(2) Cusachs, L. C., *J. Chem. Phys.* (1965) **43**, S157.
(3) Cusachs, L. C., Trus, B. L., Carroll, D. G., McGlynn, S. P., *Int. J. Quantum Chem.* (1967) 1S, 423.
(4) Cusachs, L. C., Corrington, J. H., "Sigma Molecular Orbital Theory," O. Sinanoglu and K. Wiberg, Eds., Yale University Press, New Haven, 1970.
(5) Corrington, J. H., Ph.D. Dissertation, Tulane University, 1968.
(6) Cusachs, L. C., *Spectrosc. Lett.* (1970) 3, 7.
(7) Cusachs, L. C., *Int. J. Quantum Chem.* (1967) 1S, 419.
(8) Corrington, J. H., Cusachs, L. C., *Int. J. Quantum Chem.* (1969) 3S, 207.
(9) Corrington, J. H., Aldrich, H. S., McCurdy, C. W., Cusachs, L. C., *Int. J. Quantum Chem.* (1971) **55**, 307.
(10) Wheelock, K. S., Jonassen, H. B., Cusachs, L. C., *Int. J. Quantum Chem.* (1970) 4S, 209.
(11) Cusachs, L. C., Reynolds, J. W., *J. Chem. Phys.* (1965) **43**, S160.
(12) Cusachs, L. C., Reynolds, J. W., Barnard, D., *J. Chem. Phys.* (1966) **44**, 835.
(13) Cusachs, L. C., Linn, Jr., J. R., *J. Chem. Phys.* (1967) **46**, 2919.
(14) Cusachs, L. C., Miller, D. J., McCurdy, Jr., C. W., *Spectrosc. Lett.* (1969) 2, 141.
(15) Fogleman, W. W., Miller, D. J., Jonassen, H. B., Cusachs, L. C., *Inorg. Chem.* (1969) 8, 1209.
(16) Cusachs, L. C., *Spectrosc. Lett.* (1970) 3, 195.
(17) Miller, D. J., Ph.D. Dissertation, Tulane University, 1970.
(18) Miller, D. J., Cusachs, L. C., *Chem. Phys. Lett.* (1969) 3, 501.
(19) Slater, J. C., Mann, J. B., Wilson, T. M., Wood, J. H., *Phys. Rev.* (1969) **184**, 672.
(20) Froese Fischer, C., unpublished calculations.
(21) Herman, F., Skillman, S., "Atomic Structure Calculations," Prentice Hall, Englewood Cliffs, 1963.

(22) Mann, J. B., *Los Alamos Sci. Rept.* (1967) **LA 3690, 3691.**
(23) Harrison, W. A., "Pseudopotentials in the Theory of Metals," Benjamin, New York, 1966.
(24) Hoffmann, R., *J. Chem. Phys.* (1963) **39,** 1397.
(25) Hoffmann, R., *J. Chem. Phys.* (1964) **40,** 2474, 2480.
(26) Carroll, D. G., Armstrong, A. T., McGlynn, S. P., *J. Chem. Phys.* (1966) **44,** 1865.
(27) Rein, R., Fukuda, N., Win, H., Clarke, G. A., Harris, F. E., *J. Chem. Phys.* (1966) **45,** 4743.
(28) Pople, J. A., Segal, G. A., *J. Chem. Phys.* (1965) **43,** S136.
(29) Dahl, J. P., Ballhausen, C. J., *Adv. Quantum Chem.* (1968) **4,** 170.
(30) Kaufman, J. J., *Int. J. Quantum Chem.* (1967) **1S,** 485.
(31) Hoeft, J., Lovas, F. J., Tiemanns, E., Törring, T., *J. Chem. Phys.* (1970) **53,** 2736.
(32) Siegbahn, K. *et al.,* "ESCA Applied to Free Molecules," North Holland, Amsterdam, 1969.
(33) Turner, D. W., Baker, C., Baker, A. D., Brundle, C. R., "Molecular Photoelectron Spectroscopy," Wiley, London, 1970.
(34) Miller, D. J., Ph.D. dissertation, Tulane University, 1970.
(35) Rothenburg, S., Schaefer, H. F., III, *J. Chem. Phys.* (1970) **53,** 3041.
(36) Sichel, J. M., Whitehead, M. A., *Theor. Chem. Acta.* (1968) **11,** 239.
(37) Jonassen, H. B., McCurdy, C. W., Cusachs, L. C., Ansell, G. B., Finnegan, W. G., **NWC TP 4996,** "Electronic Effects in the Structure of Some Silver Iodide Complexes," 1970.

RECEIVED March 5, 1971.

2

Electron Behavior in Some Sulfur Compounds

Vanderbilt University, Nashville, Tenn. 37203

JOHN R. VAN WAZER and ILYAS ABSAR

Ab Initio *LCAO–MO–SCF calculations have been used to elucidate the electronic structure of the molecules H_2S, H_2SO, H_2SO_2, with and without* d *orbitals being allowed to the sulfur. Three-dimensional plots of the various molecular orbitals are presented and discussed.*

During the last five years, the high capabilities of modern computers have led to the development of several programs (*1, 2, 3, 4*) for carrying out *ab initio* calculations of wave functions based on the Hartree-Fock approximation. These calculations are carried out with Gaussian or Slater-type atomic orbitals, and wavefunctions for molecules having a reasonably large number of electrons may be obtained with accuracy. Such a wavefunction gives a picture of the electronic structure of the chosen molecule which is not beclouded by using the approximations made in semiempirical work to avoid calculation of the many integrals involved in an *ab initio* study.

In this study we have carried out *ab initio* calculations on three related sulfur compounds which could be considered as representative of a large body of sulfur chemistry. Since the hydrogen atom carries only one electron, hydrogen sulfide (H_2S) was chosen as one of the three molecules, and the other two were its hypothetical oxidation products, sulfur hydrate (H_2SO) (*5*) and the sulfoxylic acid tautomer (H_2SO_2) (*5*). To study the effect of *d* character on the electronic structures, these calculations were carried out with and without *d* orbitals being assigned to the sulfur atom. Of particular interest are the short bond lengths and other multiple-bond characteristics of the sulfur–oxygen linkage, a subject which has received consideration since the early days of Valence-Bond theory (*6*).

Calculational Details

The geometry of the H_2S molecule was taken from the literature (7), where the S–H distance has been shown to be 1.328A and the HSH angle to be equal to 92.2°. Since no molecular geometry has been measured for the other two molecules, we assumed that in both of them the S–O distance is 1.42A and that the oxygen(s) and sulfur lie in a plane at right angles to the HSH plane. The OSH angle was chosen to be 106.0° in all cases, thus making the OSO angle for H_2SO_2 equal to 133.15° Using these geometries, the two hydrogens and the sulfur of all three molecules are directly superimposable as is the single oxygen of H_2SO with one of the oxygens of H_2SO_2.

The *ab initio* LCAO–MO–SCF calculations were carried out with the program MOSES (4), and the electron densities were calculated by another program (an electron-density program modified by T. H. Dunning at the California Institute of Technology and later improved by H. Marsmann and I. Absar) (8), the results from which were converted into computer-made three-dimensional plots by yet another program (9).

In the LCAO approximation the sulfur atom was described by nine *s*-type and five *p*-type Gaussian exponents [usually symbolized by (95)], while each oxygen was described by five *s*-type and two *p*-type [*i.e.* (52)], and each hydrogen by three *s*-type Gaussian exponents [*i.e.* (3)]. All of these exponents were fully optimized in SCF calculations on the ground states of the individual atoms. These atom-optimized exponents are shown in Table I. A value for the sulfur *d* exponent could not be determined from optimization of the energy of the ground-state atom, and it is well known (10, 11) that *d* orbitals obtained from excited states of atoms, in which the *d* orbitals are occupied, exhibit much larger radii than *d* orbitals which are optimized in molecules. Therefore, the extensive step of orbital optimization of the *d*-orbital exponent (which sets the orbital radius) was carried out in H_2S and in H_2SO_2.

Table I.   Gaussian Description of the Atoms

SULFUR total energy[a] = -397.1500 au
  Gaussian exponents: s type 10217, 1569.8, 366.8, 104.75, 32.91, 6.982, 2.675, 0.5561, and 0.2071; p type 100.95, 23.36, 6.942, 2.253, and 0.2977.
OXYGEN total energy = -74.2569 au
  Gaussian exponents: s type 736.6, 112.9, 26.04, 7.212, and 0.5764; p type 3.188 and 0.5486.
HYDROGEN total energy = -0.4970 au
  Gaussian exponents: s type 4.239, 0.6577, and 0.1483.

[a] When the sixfold *d*-orbital is employed, reduction to the usual fivefold spherical harmonic set also gives a 3s exponent. When this exponent (0.435) is incorporated into the atomic set, the total energy of the sulfur atom is somewhat improved, being -397.1520 a.u.

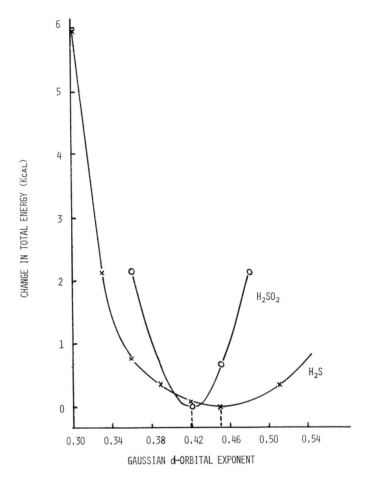

*Figure 1.   Optimization of the Gaussian d-orbital exponent in
the H₂S and the H₂SO₂ molecules*

The results for a single Gaussian d exponent optimized in each of these two molecules are given in Figure 1, which shows that the lowest energy for the molecule occurs at an orbital exponent of 0.45 for H_2S and 0.42 for H_2SO_2. Note that varying the Gaussian d-orbital exponent (a process which is equivalent to varying the orbital radius) over a wide range results in a change of only a few kcal in the total energy of the molecule. A change of 5 kcal in the total energy of the H_2S molecule represents a variation in the total energy of only 0.0020% while for the H_2SO_2 molecule it is even smaller, being 0.0015%.

Comparison of Calculations on Hydrogen Sulfide. Our calculated molecular energy for H_2S in the (951/3) basis set is -398.3600 au. When $3d$ orbitals were omitted—*i.e.*, in the (95/3) GTO basis set—the total molecular energy was -398.3117 au. The notation $(abc/ef/g)$ refers to the use of a s-type, b p-type, and c d-type Gaussian exponents for the sulfur atom; e s-type and f p-type for each oxygen; and g s-type for each hydrogen atom. For H_2S, the symbolism is (abc/g).

At least six *ab initio* calculations for H_2S have been reported previously (*12, 13, 14, 15, 16, 17*). Table II compares the present SCF energies with those obtained by previous calculations and with the results of a semiempirical CNDO calculation (*18*). From a comparison of the different results, it seems that our uncontracted Gaussian basis set calculation gives a lower total energy than any of the Slater-type orbital calculations. However, the large Gaussian basis set, (12 9 1/51) (*16*), gives about 0.3 au better energy than does our (951/3). The other Gaussian results (*14, 17*) do not give energies as low as those reported here.

The calculations of Boer and Lipscomb (*13*) are of particular interest because they have also studied the effect of allowing d character on the sulfur atom using a minimum basis set of Slater-type functions as opposed to our uncontracted Gaussian basis set. They give a value of -397.7881 au for the molecular energy of H_2S and report a decrease to -397.8415 au when a molecularly optimized $3d$ orbital with a 1.7077 exponent is included with the minimum Slater basis set. From this Slater exponent the radius for the shell of maximum electron probability of the sulfur $3d$ atomic orbitals is 1.76 A while, from our molecularly optimized Gaussian $3d$ exponent, this radius is 1.83 A. A semiempirical study (*19*) using Slater functions gave a corresponding radius of 1.79 A. This observation implies a correspondence between our Gaussian-basis calculation and the Slater-basis calculation of Boer and Lipscomb. If d character is allowed, both the calculations report an improvement of 0.05 au in the molecular energies. Also, as Boer and Lipscomb noted (*13*), it is apparent from our calculations that when d orbitals are included, the gross electronic populations (*20*) of the atoms are built up mainly at the expense of the hydrogen $1s$ atomic orbitals rather than the outer-shell atomic orbitals of the sulfur.

The calculated energy needed to dissociate the molecule into atoms (*i.e.*, the binding energy) from the (951/3) basis set is 5.822 ev while that from the (95/3) basis set is 4.565 ev. Note that the experimental value obtained from $\Delta H = -4.232$ kcal/mole (*21*) at $0°K$ for the formation of H_2S (g) from the ground-state gaseous atoms is 7.926 ev, using a correction of 9.2 kcal/mole for the zero-point vibrational energy obtained from the fundamental frequencies listed by Herzberg (*22*). There seems to be some confusion in the literature as to the experimental value of the binding energy. Rothenberg *et al.* (*16*) report a value of 6.96 by adding the H–SH dissociation energy (3.26 ev) to the dissociation energy

Table II.   Orbital Energies and Total Energies

	1a$_1$	2a$_1$	1b$_2$	3a$_1$
			Molecular-Orbital	
United atom (12)				
One-center STO (15)	−92.479	−9.080	−6.636	−6.635
Minimum-basis STO (13)				
Minimum-basis STO with 3d (13)	−91.902	−8.775	−6.447	−6.446
STO using Gaussian integrals (14)				
52/1 GTO uncontracted (17)	−91.857	−8.645	−4.520	−4.519
521/1 GTO uncontracted (17)	−92.023	−8.850	−4.428	−4.427
12 9 2/51 GTO contracted (16)	−91.974	−8.962	−6.652	−6.651
This work				
95/3 GTO uncontracted	−91.992	−9.003	−6.672	−6.670
951/3 GTO uncontracted	−91.979	−8.987	−6.657	−6.656
CNDO				

of SH (3.70 ev), however, without taking into account the zero-point vibrational correction. Boer and Lipscomb (13) report 7.031 ev but do not indicate their source for this value. If their value were obtained from thermodynamic data like ours, there must have been errors in calculation.

Table III presents our calculated values of the molecular ground-state properties for H_2S and the results of a CNDO calculation are included also. The calculated binding energies for hydrogen sulfide are smaller than the observed value. This finding is attributed to the molecular extra-correlation energy, which is often one-quarter to one-third of the value of the binding energy calculated from the SCF approximation at the Hartree–Fock limit (23). The calculated binding energies given in Table III for hydrogen sulfide appear to exhibit about the expected value. Also, our values of binding energy, first ionization potential, and dipole moment compare favorably with the other previous calculations on H_2S.

Even though the binding energy obtained from the CNDO calculation appears near the experimental value, it is too large when account is taken of the necessary correlation correction. The value of the ionization potential obtained from CNDO is also too large while the dipole moment is much too small. We have noticed (24, 25) that properties calculated from CNDO show no uniform behavior. In one instance the calculated dipole moment agrees with the experimental value, while in another instance the agreement is extremely poor. This statement is also true of ionization potentials and binding energies. An examination of the valence-shell orbital energies shown in Table II indicates that those calculated by the CNDO method for H_2S are consistently lower than the orbital energies obtained from any of the *ab initio* calculations. Therefore, it seems that the semiempirical CNDO method does not give results comparable with

of H_2S and S in Their Electronic Ground States

Energy, au					*Total Molecular*
$1b_1$	$4a_1$	$2b_2$	$5a_1$	$2b_1$	*Energy, au*
					−394.6794
−6.632	−0.940	−0.531	−0.454	−0.351	−397.5891
					−397.7881
−6.442	−0.935	−0.561	−0.466	−0.346	−397.8415
					−396.9349
−4.516	−0.797	−0.422	−0.232	−0.125	−379.6370
−4.425	−0.810	−0.417	−0.242	−0.120	−381.0389
−6.649	−0.986	−0.594	−0.501	−0.383	−398.6862
−6.667	−0.986	−0.555	−0.456	−0.341	−398.3117
−6.653	−0.966	−0.555	−0.462	−0.329	−398.3600
	−1.001	−0.640	−0.590	−0.491	

ab initio calculations. We conclude that the erratic trends in properties obtained from CNDO calculations render the method quite unreliable. This fact is not surprising, considering the sweeping approximation involved in the evaluation of integrals by the CNDO method.

From the energy and calculated electronic properties, it appears that for H_2S both the (95/3) and (951/3) basis sets are superior approximations than is a Slater minimum basis set. It should also be pointed out that calculations using large uncontracted Gaussian basis sets are sensitive to contraction procedures, while too small an uncontracted Gaussian basis set (17) is inadequate. A useful uncontracted Gaussian basis set should be at least sufficiently large so as to give results comparable with those obtained from a minimum Slater basis set.

Comparison between H_2S, H_2SO, and H_2SO_2—Molecular Properties. The binding energies shown in Table III demonstrate the great influence of adding d character to the sulfur when oxygen atoms are present. Thus, the binding energy for H_2S is increased in our calculations by only 1.26 ev upon allowing d orbitals. The equivalent numbers for the H_2SO and H_2SO_2 molecules are 5.12 and 9.07 ev, respectively. These values correspond to an added stabilization of the molecule by permitting d character to the sulfur at 3.9 ev per added oxygen atom. With d orbitals being disallowed, the H_2SO_2 molecule is unstable with respect to the constituent atoms. However, allowing d character to the sulfur causes all three of the molecules studied here to exhibit about the same stability with respect to their ground-state atoms.

In Table III, the results for calculations with and without d orbitals on the sulfur are intercompared. Allowing d orbitals to the sulfur caused a decrease of 0.02% in the total energy for H_2S, 0.04% for H_2SO, and 0.06% for H_2SO_2. As expected, the value of the dipole moment of H_2S

was reduced when d orbitals were incorporated into the description of the molecule. The sign of the dipole corresponds to the hydrogens being at the positive end of the C_{2v} axis of the molecule.

In all cases, allowing d character to the sulfur resulted in a transfer of electronic charge from the hydrogens and from the oxygen(s), if present, to the sulfur. The most pronounced charge transfer was, as expected (24, 26, 27), from oxygen to sulfur, amounting to a transfer of about a third of an electron per S–O bond, presumably arising from addition of p_π–d_π bonding character. Although the calculation of atomic charges is an arbitrary procedure (25, 28), the values for these formal quantities given in Table III are valuable for purposes of comparison. Thus, for the case in which d orbitals are supplied to the sulfur atom, when adding the first oxygen atom to hydrogen sulfide, there is a large decrease of around 0.4 electron in the charge on the sulfur atom; addition of the

Table III. Calculated Molecular Ground-State Properties

	Basis Set		
Property	(95/52/3)	(951/52/3)[a]	CNDO
Hydrogen sulfide, H_2S	−398.3116	−398.3600	
total energy, au			
binding energy, eV	−4.564	−5.822	−7.828
observed value	−7.926	−7.926	−7.926
dipole moment, D	1.301	0.725	0.071
observed value	1.019	1.019	1.019
charge on H atom	+0.046	+0.101	+0.004
charge on S atom	−0.093	−0.202	−0.007
Sulfur hydrate, H_2SO	−472.4064	−472.5966	
total energy, au			
binding energy, eV	−0.1521	−5.272	
dipole moment, D	3.167		
charge on H atom	−0.006	+0.073	
charge on S atom	+0.662	+0.201	
charge on O atom	−0.651	−0.347	
Sulfoxylic acid tautomer, H_2SO_2	−546.5526	−546.8880	
total energy, au			
binding energy, eV	+2.862	−6.214	
dipole moment, D	2.631		
charge on H atom	−0.0001	+0.109	
charge on S atom	+1.250	+0.333	
charge on O atom	−0.625	−0.276	

[a] The d-orbital exponent chosen to be 0.45, 0.435, and 0.42 for H_2S, H_2SO, and H_2SO_2 respectively. The symbols (95/52/3) and (951/52/3) mean that 9 s-type, 5 p-type and respectively either none or 1 d-type orbital exponents are used to describe the sulfur atom; 5 s-type and 2 p-type to describe each oxygen, if present; and 3 s-type to describe each hydrogen atom.

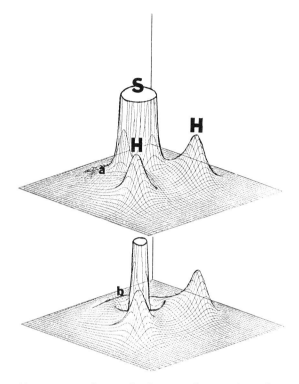

Figure 2. The total electron density (top diagram) and valence orbital electron density (bottom diagram) in the H $\overset{S}{\diagup\diagdown}$ *H plane of the* H_2SO_2 *molecule. The molecular geometry is shown on the basal plane of each diagram and the electron density perpendicular to it.*

second oxygen leads to an additional increase in charge of only about 0.1 electron.

Electron–density diagrams are shown in Figure 2 as measured across the plane passing through the sulfur and the two hydrogen atoms and in Figure 3, as measured across the perpendicular plane passing through the sulfur and bisecting the H–S–H angle, the plane which contains the oxygen atoms, when present. In these figures and all of the following, the geometry of the molecule is shown on the base plane while the intensity of the electrons in that plane is depicted perpendicular to it.

In Figure 2, the total and valence-orbital electron densities in the H $\overset{S}{\diagup\diagdown}$ H plane are shown for the H_2SO_2 molecule only, since the H_2S and H_2SO molecules exhibit similar plots. The main part of the total electron density in this plane is centered around the sulfur nucleus, where the electron density rises so high that its peak had to be cut off near the base. The plot for the valence orbitals shows a central peak at the sulfur surrounded by a ring of charge as demanded by the necessity for the valence shell to be orthogonal with the inner orbitals on the sulfur. The difference between the valence-shell plot for the H_2SO_2 and H_2S molecules is very slight, primarily showing up for H_2S in the region behind the hydrogens as a somewhat greater height of the low "hill" of electron density appearing around the sulfur atom (*see* the region labeled a in Figure 2). For the valence orbitals, the difference shows up in the same place and corresponds to an increase in electron density for the H_2S as compared with the H_2SO_2 molecule in the region labeled b in Figure 2.

Since the electron-density plots for these molecules are quite different in the plane passing through the sulfur at right angles to the $H \overset{\diagup S \diagdown}{\quad} H$ plane and bisecting the H–S–H angle, they are shown in Figure 3 for all three of the molecules studied. The electronic effect of the sequential replacement of the unshared pairs of electrons on the sulfur by oxygen is seen most clearly in the electron-density plots for the valence-orbital shells which also are shown in Figure 3.

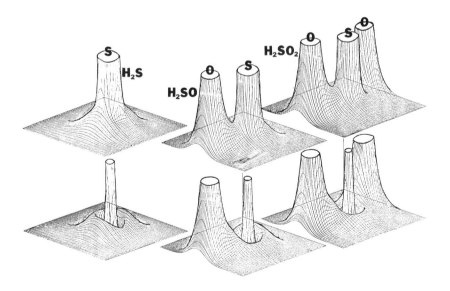

Figure 3.   The total electron densities (top diagrams) and valence orbital electron densities (bottom diagrams) in the plane perpendicular to the $H \overset{\diagup S \diagdown}{\quad} H$ plane passing through the sulfur atom and bisecting the H–S–H angle for the three molecules: H_2S, H_2SO, and H_2SO_2

Table IV.   Molecular-Orbital Properties of H_2S

Electronic Populations[a]

Orbital	Energy (ev)	H gross	S gross	S–H overlap per bond	Major Contribution[b]
$1a_1$	−2502.7621	0.000	2.000	0.000	S_{1s}
Δ[c]	+0.3388	0.000	0.000	0.000	
$2a_1$	−244.5367	−0.000	2.001	−0.001	S_{2s}
Δ	+0.4260	−0.001	0.002	−0.002	
$1b_2$	−181.1331	−0.000	2.001	−0.001	S_{2p}
Δ	+0.4055	0.000	0.000	−0.001	
$3a_1$	−181.1098	−0.000	2.001	−0.001	S_{2p}
Δ	+0.3844	0.000	0.000	−0.002	
$1b_1$	−181.0219	−0.000	2.000	−0.000	S_{2p}
Δ	+0.3849	0.000	0.000	0.000	
$4a_1$	−26.2881	0.186	1.629	0.236	S–H
Δ	+0.5481	0.006	−0.011	0.007	
$2b_2$	−15.0892	0.471	1.058	0.384	S–H
Δ	+0.0011	−0.064	0.128	0.033	
$5a_1$	−12.5803	0.243	1.513	−0.091	S_{lp}
Δ	−0.1600	0.005	−0.011	−0.002	
$2b_1$	−8.9579	0.000	2.000	−0.000	S_{lp}
Δ	+0.3084	0.000	0.000	0.000	
Total		0.899	16.202	0.527	
Δ		−0.055	0.109	0.036	

[a] *See* Ref. *28* for definitions.
[b] This is the dominant valence-bond contribution to the delocalized orbital—the contribution which would be found to make up the major part of the molecular orbital if it were subjected to proper delocalization.
[c] The Δ value in italics shows the effect of adding d character to the sulfur. It gives the difference between the value for each entry as calculated in the (951/52/3) basis set minus that calculated in the (95/52/3) set.

Orbital Properties

Tables IV through VI give the orbital symmetries, their energies, and the electronic populations, as well as the major bonding contribution to each of the delocalized orbitals for the molecules H_2S, H_2SO, and H_2SO_2 respectively. The numbers in italics in these tables (numbers which are labeled Δ) show the change, induced by allowing d character, in the particular value associated with the given orbital. For example, the value of Δ of 0.155 presented in Table V for the S–O overlap population of orbital $3a''$ of the H_2SO molecule indicates that there is an increase of 0.155 electrons in the S–O bond when d character is allowed—i.e., the calculated value of the S–O overlap population of orbital $3a''$ of H_2SO for

Table V.   Molecular-Orbital Properties of H_2SO

Orbital	Energy (ev)	gross H	gross S	gross O	overlap per bond S–H	overlap per bond S–O	Major Contribution[b]
1a′	−2507.0699	0.000	2.000	0.000	0.000	0.000	S_{1s}
Δ[c]	*0.9243*	*0.000*	*0.000*	*0.000*	*0.000*	*0.000*	
2a′	−560.6314	0.000	−.004	2.004	0.000	−0.009	O_{1s}
Δ	*−1.8269*	*0.000*	*−.002*	*.003*	*0.000*	*−0.006*	
3a′	−248.4883	0.000	2.000	0.000	−0.001	0.001	S_{2s}
Δ	*1.2886*	*−0.001*	*.002*	*0.000*	*−0.002*	*0.001*	
4a′	−185.1021	0.000	2.001	0.000	0.000	−0.001	S_{2p}
Δ	*1.2420*	*0.000*	*.001*	*−0.001*	*−0.001*	*−0.001*	
1a″	−185.0316	0.000	2.001	0.000	−0.001	0.000	S_{2p}
Δ	*1.2326*	*−0.001*	*0.000*	*0.000*	*0.000*	*0.000*	
5a′	−184.9851	0.000	2.000	0.000	0.000	0.000	S_{2p}
Δ	*1.2167*	*0.000*	*0.000*	*0.000*	*0.000*	*0.000*	
6a′	−37.6247	0.012	0.730	1.245	0.016	0.614	S–O
Δ	*0.1824*	*0.002*	*−.022*	*0.018*	*0.001*	*0.047*	
7a′	−25.7925	0.194	1.249	0.364	0.236	0.006	S–H
Δ	*0.6256*	*0.023*	*0.042*	*−0.089*	*0.024*	*0.129*	
2a″	−17.5594	0.314	1.131	0.241	0.299	0.129	S–H
Δ	*0.3236*	*0.036*	*0.030*	*0.042*	*0.005*	*0.014*	
8a′	−15.8779	0.098	1.074	0.731	0.031	0.162	S_{lp}
Δ	*0.1266*	*−0.001*	*−0.205*	*0.207*	*0.007*	*−0.011*	
9a′	−13.4956	0.003	0.743	1.251	0.004	0.121	O_{lp}
Δ	*−0.7531*	*−0.015*	*0.050*	*−0.020*	*−0.003*	*0.011*	
3a″	−10.1906	0.181	0.252	1.387	0.055	0.198	O_{lp}
Δ	*−1.1538*	*−0.041*	*0.258*	*−0.176*	*0.020*	*0.155*	
10a′	−8.7918	0.127	0.622	1.124	−0.172	−0.038	O_{lp}
Δ	*−0.6559*	*−0.009*	*0.308*	*−0.290*	*−0.012*	*0.137*	
Total		0.927	15.799	8.347	0.467	1.183	
Δ		*−0.078*	*0.461*	*−0.305*	*0.135*	*0.597*	

[a] *See* Ref. *28* for definitions.
[b] This is the dominant valence-bond contribution to the delocalized orbital—the contribution which would be found to make up the major part of the molecular orbital if it were subjected to proper delocalization.
[c] The Δ value in italics shows the effect of adding d character to the sulfur. It gives the difference between the value for each entry as calculated in the (951/52/3) basis set minus that calculated in the (95/52/3) set.

Table VI   Molecular-Orbital Properties of H_2SO_2

Or-bital	Energy (ev)	Electronic Populations[a] gross			overlap per bond		Major Contri-bution[b]
		H	S	O	S–H	S–O	
$1a_1$	−2510.4142	0.000	2.000	0.000	0.000	0.000	S_{1s}
Δ^c	1.8640	0.000	0.000	0.000	0.000	0.000	
$1b_1$	−562.4887	0.000	−0.004	1.002	0.000	−0.005	O_{1s}
Δ	−1.7007	0.000	−0.003	0.001	0.000	−0.003	
$2a_1$	−562.4851	0.000	−0.004	1.002	0.000	−0.005	O_{1s}
Δ	−1.7004	0.000	−0.003	0.001	0.000	−0.003	
$3a_1$	−251.3738	0.000	2.000	0.000	−0.001	0.001	S_{2s}
Δ	2.4847	−0.001	0.002	0.000	−0.003	0.001	
$2b_1$	−188.0794	0.000	2.000	0.000	0.000	0.000	S_{2p}
Δ	2.4032	0.000	0.000	0.000	0.000	−0.001	
$4a_1$	−187.9885	0.000	2.001	0.000	−0.001	0.000	S_{2p}
Δ	2.3879	0.000	0.000	0.000	−0.001	0.000	
$1b_2$	−187.9837	0.000	2.001	0.000	−0.001	0.000	S_{2p}
Δ	2.3602	0.000	0.000	0.000	−0.001	0.000	
$5a_1$	−40.3096	0.018	0.784	0.590	0.013	0.010	S–O
Δ	0.9663	0.005	−0.065	0.027	0.013	0.010	
$3b_1$	−37.1928	0.000	0.510	0.745	0.000	0.316	S–O
Δ	0.2017	0.000	0.061	−0.031	0.000	0.035	
$6a_1$	−25.1471	0.220	0.986	0.287	0.248	0.009	S–H
Δ	1.1118	0.031	−0.018	−0.023	0.018	0.106	
$2b_2$	−19.3964	0.232	1.147	0.195	0.240	0.109	S–H
Δ	1.0750	−0.012	−0.050	0.037	0.001	0.004	
$7a_1$	−17.7018	0.040	0.778	0.571	0.039	0.091	O_{lp}, S_{lp}
Δ	0.1092	0.004	0.000	−0.004	0.026	0.075	
$4b_1$	−15.4336	0.000	0.389	0.806	0.000	0.009	O_{lp}
Δ	−0.5469	0.000	−0.074	0.037	0.000	0.094	
$1a_2$	−12.4471	0.000	0.263	0.868	0.000	0.159	O_{lp}
Δ	−1.2452	0.000	0.263	−0.132	0.000	0.159	
$8a_1$	−12.1631	0.093	0.356	0.729	0.060	0.145	O_{lp}
Δ	−1.2430	−0.069	0.335	−0.098	0.077	0.121	
$5b_1$	−11.5885	0.000	0.205	0.898	0.000	0.113	O_{lp}
Δ	−1.0781	0.000	0.203	−0.101	0.000	0.113	
$3b_2$	−10.7142	0.288	0.257	0.584	0.106	0.054	O_{lp}
Δ	0.4264	−0.069	0.264	−0.064	0.137	0.034	
Total		0.890	15.668	8.276	0.711	1.305	
Δ		−0.110	0.917	−0.349	0.259	0.777	

[a] *See* Ref. *28* for definitions.
[b] This is the dominant valence-bond contribution to the delocalized orbital—the contribution which would be found to make up the major part of the molecular orbital if it were subjected to proper delocalization.
[c] The Δ value in italics shows the effect of adding d character to the sulfur. It gives the difference between the value for each entry as calculated in the (951/52/3) basis set minus that calculated in the (95/52/3) set.

the (951/52/3) basis set is 0.198 electron, as shown in Table IV whereas the same value for the (95/52/3) is 0.043 electrons.

From the populations as well as from the orbital energies shown in Table IV, we see that orbitals $1a_1$, $2a_1$, $1b_2$, $3a_1$, and $1b_1$ of the H_2S molecule represent the $1s$, $2s$, $2p_y$, $2p_x$, and $2p_z$ inner-shell orbitals of the sulfur atom which have been subjected to a small amount of delocalization and distortion because of their presence in the molecule rather than in the isolated atom. This delocalization is barely picked up by the elec-

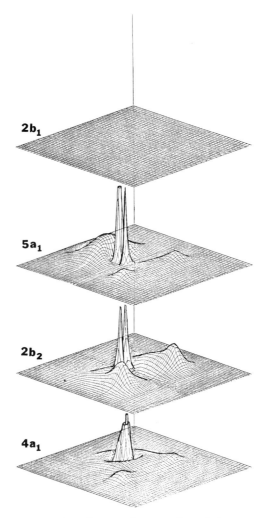

Figure 4.   Electron-density plots of the valence-shell molecular orbitals of the H_2S molecule in the H⌢S⌢H plane

tronic population as reported in these tables to the third decimal place. Thus, for orbitals $2a_1$, $1b_1$, and $3a_1$ of H_2S there is a donation of 0.001 electron to the sulfur from the pair of hydrogen atoms.

The energies of the $1s$, $2s$, and $2p$ orbitals of the free sulfur atom are -2502.1271 ev, -243.9804 ev, and -180.4858 ev, respectively. These values should be compared with the similarly assigned inner orbitals of Tables IV through VI which show there are changes in the inner-orbital energies when going from the atom to the various sulfur-containing molecules. Such changes are related to the study (*29, 30*) of inner-orbital binding energies by photoelectron spectroscopy—a subject which has achieved considerable interest among chemists during the past few years.

Returning to Table IV, there are four, filled, more or less delocalized, valence-shell molecular orbitals for the H_2S molecule, as expected. The two more stable of these orbitals, $4a_1$ and $2b_2$, are seen from the electronic populations to be dominated by S–H bonding whereas the less stable pair, $5a_1$ and $2b_1$, exhibit major contributions from the lone-pair electrons on the sulfur, with some delocalization of orbital $5a_1$ onto the hydrogen atoms, coupled with a small amount of S–H antibonding.

Electron-density plots for each valence orbital of H_2S are shown in Figure 4 for the H $\overset{\diagup S\diagdown}{}$ H plane and in Figure 5 for the plane, at right angles to this one, passing through the sulfur atom and bisecting the H–S–H angle. Since the node for orbitals of b_1 symmetry is in the H $\overset{\diagup S\diagdown}{}$ H plane, the plot for the $2b_1$ orbital in Figure 4 shows no electron density. A similar situation holds for orbitals of b_2 symmetry (as demonstrated by orbital $2b_2$ in Figure 5) with respect to the plane passing through the sulfur atom at right angles to the H $\overset{\diagup S\diagdown}{}$ H plane.

From Figures 4 and 5 we note that orbitals $4a_1$ and $2b_2$ represent S–H bonding whereas orbitals $5a_1$ and $2b_1$ account for the unshared electrons on the sulfur atom. These conclusions agree with the assignments made previously in Table IV using the electronic populations.

Analyzing the molecular valence-orbital plots is facilitated by comparing them with equivalent plots for the atomic valence orbitals of the sulfur, hydrogen, and the oxygen atoms in Figure 6. It is clear by comparing Figure 6 with Figures 4 and 5 that the $4a_1$ orbital of the H_2S molecule is dominated by sulfur s character whereas the three other valence orbitals of H_2S are dominated by sulfur p character. Orbitals $4a_1$ and $2b_2$ in Figure 4 clearly indicate that bonding to a second-row atom such as sulfur involves the diffuse outer antinode of the atom. Such bonds never show as great a localized concentration of electrons in the

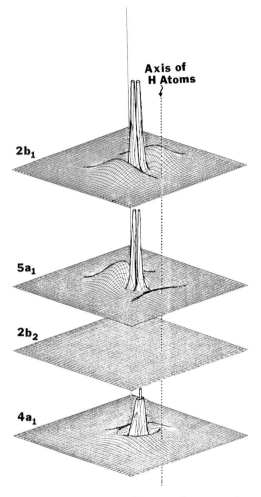

Figure 5. Electron-density plots for the valence-shell molecular orbitals of the H_2S molecule in the plane perpendicular to the

$$\overset{S}{\underset{H\qquad H}{\diagup\diagdown}}$$

plane passing through the sulfur atom and bisecting the H–S–H angle

bonding region as found in an analogous structure based on the first-row atom from the same group (20). However, the overlap population may be about the same, since the bonding to the second-row atom corresponds to the electrons being spread out in space (31). Orbitals $4a_1$, $2b_2$, and $5a_1$ of Figure 4 also show that there is a concentration of electronic charge near the sulfur nuclei for these valence-shell orbitals, owing to the anti-node(s) lying inside of the outer one which is involved in the electron sharing between the sulfur and the hydrogens.

We rationalize delocalization found in molecules of second-row atoms in terms of the diffuseness of the outer antinodes. It is illustrated by orbital $5a_1$ in Figure 4, where it might be argued that the presence of the two protons on the orbital which accounts for the lone-pair electrons of the sulfur inadvertently leads to charge concentrations at the positions of the two hydrogen atoms.

The six orbitals listed towards the top of Table V represent the inner-shell molecular orbitals of sulfur and oxygen for the H_2SO molecule. Note that the inner-shell orbital of oxygen showing predominately $1s$ character lies at a considerable lower energy than the sulfur "$2s$" and "$2p$" orbitals as for the free atoms. The remaining seven molecular orbitals of H_2SO correspond to the valence shell, with the orbital dominated by S–O bonding being more stable than the pair of orbitals dominated by S–H bonding and the four orbitals predominately representing the lone pairs being the least stable. These assignments of the various molecular orbitals of H_2SO are borne out by their electron density plots shown in Figures 7 and 8 where orbital $6a'$ represents what might be called σ bonding between the sulfur and the oxygen, with orbitals $7a'$ and $2a''$ being dominated by S–H bonding as predicted in Table V by analyzing the population. Note the polarity towards the oxygen of the S–O bond shown in orbital $6a'$ of Figure 8.

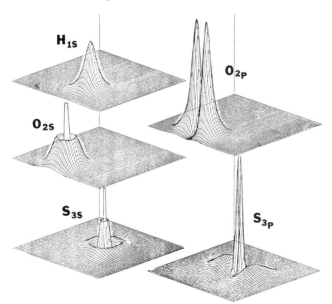

Figure 6. *Electron-density plots for the valence-shell atomic orbitals of sulfur, oxygen, and hydrogen as the free atoms. These plots are to the same scale as those given in Figures 2 through 5 and 7 through 10.*

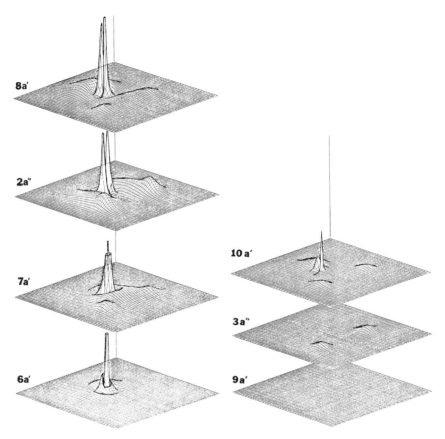

Figure 7.   Electron-density plots of the valence-shell molecular orbitals of the
H_2SO *molecule in the H* ⁀S⁀ *H plane*

Our assignment in Table IV of the sulfur lone-pair as being the major contribution to orbital 8a′ is interesting because of the highly delocalized nature of this orbital. Figure 7 shows (in agreement with Table V) that there is a small amount of S–H bonding with a reasonable electron density on the two hydrogens and in the region between them. Figure 7 also shows a pile-up of charge on the side of the sulfur atom opposite to the hydrogens in the H ⁀S⁀ H plane. The plane at right angles to this one containing the oxygen and sulfur atoms (*see* orbital 8a′ in Figure 8) exhibits a small amount of what might be called S–O π bonding, as well as a considerable pile-up of electrons around the oxygen nucleus in orbital 8a′.

The predominance of the oxygen lone pairs in orbitals 9a′, 3a″, and 10a′ of the H_2SO molecule shows up in Figure 8 and is substantiated by

Figure 7. In Table V, it appears from the S–O overlap population that orbitals $9a'$ and $3a''$ exhibit S–O bonding whereas $10a'$ shows S–O anti-bonding. In spite of the fact that p character dominates the oxygen and sulfur contributions to orbital $9a'$ (as can be seen in Figure 8), any S–O bonding in this molecular orbital must be considered as σ-type bonding because of the orientations of the respective p-type nodal planes of the sulfur and of the oxygen. Based on the same type of argument, the S–O bonding of orbital $3a''$ must be considered as π-like, and this is also true for the antibonding character of orbital $10a'$. It is clear from the information afforded by the electron-density plots of Figure 8 that the relatively large values of Δ for the S–O overlap of orbitals $3a''$ and $10a'$,

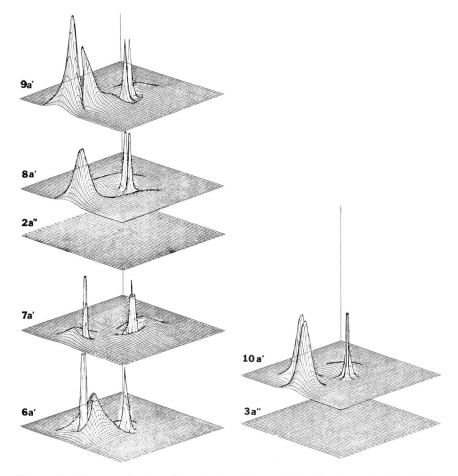

Figure 8. Electron-density plots of the valence-shell molecular orbitals of the H_2SO molecule in the plane containing the oxygen and the sulfur atoms, the plane which bisects the H–S–H angle

38   SULFUR RESEARCH TRENDS

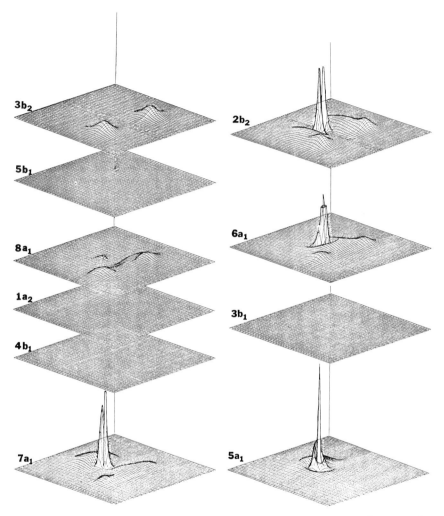

Figure 9.   Electron-density plots of the valence-shell molecular orbitals of the
H_2SO_2 *molecule in the* $H\overset{\diagup S\diagdown}{}H$ *plane*

shown in Table V, correspond to the introduction of p_π–d_π feedback between the oxygen and sulfur atoms when d character is allowed to the sulfur. Electron-density difference-plots not shown here demonstrate that upon allowing d character to the sulfur atom, there is a transfer of charge from the regions on either side of the oxygen atom corresponding to oxygen lone-pair electrons to a region around the S–O bond where the electron density exhibits π symmetry with respect to that bond.

As evidenced by their respective energies and electronic populations, the seven orbitals listed toward the top of Table VI are the inner-shell molecular orbitals of sulfur and oxygen in the H_2SO_2 molecule. Of the remaining ten molecular orbitals making up the valence shell (*see* Table VI), the two more stable ones are attributable primarily to the S–O bonds with the next two being dominated by the S–H bonds. Each of the six less energetic orbitals exhibit major contributions caused by the oxygen lone pairs. Electron-density plots for each of the valence-shell molecular orbitals of H_2SO_2 are shown for the $\overset{\diagup S \diagdown}{H \qquad H}$ plane in Figure 9 and for the $\overset{\diagup S \diagdown}{O \qquad O}$ plane in Figure 10. Note that the plots for orbitals $5a_1$ and $3b_1$ in in Figure 10 each correspond to a pair of highly polar S–O bonds, all being of the σ-type, with sulfur s character predominating in $5a_1$ and p character in $3b_1$. The S–H bonding shows up in orbitals $6a_1$ and $2b_2$ of Figure 9, with sulfur s character dominating $6a_1$ and sulfur p character dominating $2b_2$. The electron-density plot for orbital $6a_1$ in Figure 10 demonstrates that the gross charge on the oxygen of 0.287 electron shown in Table VI for this orbital corresponds to lone pairs on the oxygen. The hump towards the front of the representation of orbital $6a_1$ in Figure 10 corresponds to the significant electron density between the hydrogen atoms which shows up more clearly for this orbital in Figure 9.

What we have designated in Table VI as a sulfur lone pair in orbital $7a_1$ is seen clearly in the electron density plot of this orbital in Figure 10. The electron density plots of Figure 10 demonstrate the dominance of oxygen lone pairs in orbitals $7a_1$, $4b_1$, $8a_1$, and $5b_1$ of the H_2SO_2 molecule. For all of these molecular orbitals, only the nodal planes of the oxygen atoms in orbitals $8a_1$ and $5b_1$ are properly directed to allow for p_π–d_π feedback from the oxygens to the sulfur, in the situation where d character is afforded to the sulfur atom. This concept of π-bond feedback is supported by the large values of Δ for the S–O overlap shown for these two orbitals in Table VI. Presumably the same situation holds for orbital $1a_2$, but its symmetry is such that it cannot be seen in Figure 10 since the nodal plane of the a_2 orbital coincides with the $\overset{\diagup S \diagdown}{O \qquad O}$ plane.

Orbital Similarity between Molecules

Figures 4 through 10 show that certain orbitals in the three different molecules H_2S, H_2SO, and H_2SO_2 are similar. This is true for each of the two molecular orbitals which account in great part for S–H bonding—*e.g.*, we find that orbital $4a_1$ in Figure 4, $7a'$ in Figure 7, and $6a_1$ in Figure 9 are similar, while orbital $2b_2$ in Figure 4 is practically indistinguishable from $2a''$ in Figure 7 and $2b_2$ in Figure 9. Thus, we see that in the $\overset{\diagup S \diagdown}{H \qquad H}$ plane, each of the pair of molecular orbitals corresponding to the S–H bond remains practically unchanged from molecule to molecule in this series of compounds! Looking at these two sequences of orbitals ($4a_1$ vs. $7a'$ vs. $6a_1$, and $2b_2$ vs. $2b'$ vs. $2b_2$) in the plane passing through the sulfur at right angles to the H$\overset{\diagup S \diagdown}{}$H plane, we see the same thing.

Similarities such as the ones described in the above paragraph may be extended readily to the orbitals which primarily describe S–O bonding as well as to the orbitals which are dominated by lone pairs of electrons.

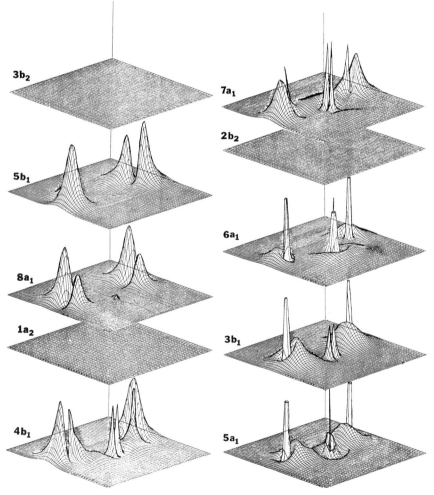

Figure 10.   Electron-density plots of the valence-shell molecular orbitals of the H_2SO_2 molecule in the O$\overset{\diagup S \diagdown}{}$O plane

Table VII. Correlation Between Orbitals in the Different Molecules

Major Contribution	Orbital in		
	H_2S	H_2SO	H_2SO_2
S–H	$4a_1$	$7a'$	$6a_1$
	$2b_2$	$2a''$	$2b_2$
S–O		$6a'$	$5a_1$
lone pairs	$5a_1$	$8a'$	$7a_1$
	$2b_1$	$9a'$	$4b_1$
		$10a'$	$8a_1$
		$3a''$	$3b_1$

A correlation table setting forth these relationships is presented as Table VII.

Of particular interest in Table VII is the sequence of orbitals $2b_1$, $9a'$, and $4b_1$ as well as the sequence of $5a_1$, $8a'$, and $7a_1$, since these sequences involve a switch from the sulfur lone pair to the oxygen lone pair when progressing along the series corresponding to increasing oxygenation of hydrogen sulfide. The correctness of placing the orbitals in the same category for each set of three is made clear by inspecting Figures 5, 8, and 10. For the $5a_1$, $8a'$, and $7a_1$ orbitals, the reader also should inspect Figures 4, 7, and 9.

Conclusions

We hope we have demonstrated that the usual delocalized molecular orbitals obtained from self-consistent-field calculations are about as readily understood and interpretable in chemical terms as are the localized orbitals which some people have taken great pains to derive (32) from these delocalized ones. We have demonstrated also through the plots of electron density how these delocalized molecular orbitals are made up from atomic orbitals and how they may be discussed in terms of their atomic-orbital composition. This is particularly interesting for second-row atoms such as sulfur because of the diffuse nature of the parts of the valence-shell atomic orbitals which are involved in bonding.

It has been shown also that certain delocalized orbitals appearing in one molecule reappear with hardly any changes in another related molecule. Presumably, there are a relatively limited number of Hartree–Fock orbitals which appear over and over again in common chemical compounds. It should be interesting to identify these "molecularly invariant" orbitals so that their properties, such as the respective electron-binding energy, may be compared to give a better insight into the electronic structure of matter.

Although much has been made of the fact that through unitary transformations a given set of molecular orbitals which describe a molecule may be converted to other sets (which may be more or less "localized" than the original set), this is not to say that the Hartree–Fock molecular orbitals described here have no special significance as do the usual s, p, d, etc. atomic orbitals. Treatment of an atom in terms of s, p, d, etc. is also neither foreordained nor necessary. However, whatever extent one chooses to argue for the usual s, p, d atomic orbitals as opposed to some other method of describing the atom, so one must attribute equal significance to the Hartree-Fock molecular orbitals. Therefore, we claim that the electron density diagrams of the molecular orbitals in Figures 4, 5, and 7–10 have as much meaning (in the same sense) as do electron density plots for atomic orbitals (*see* Figure 6).

Acknowledgment

We thank the National Science Foundation for partial support of this work.

Literature Cited

(1)   Csizmadia, I. G., Harrison, M. C., Moskowitz, J. W., Seung, S., Sutcliffe, B. T., Barnett, M. P., *Theor. Chim. Acta (Berlin)* (1966) **6**, 191.
(2)   Davis, D. R., Clementi, E., *J. Comput. Phys.* (1966) **1**, 223.
(3)   McLean, A. D., Yoshimine, M., *IBM J. Res. Develop.* (1968) **12**, 206.
(4)   Sachs, L. M., Geller, M., *Int. J. Quantum Chem.* (1969) **1S**, 445.
(5)   "Gmelin's Handbuch der Anorganischen Chemie," Vol. 9, Part B, Sec. 2, p. S[B] 374, p. S[B] 383, Verlag Chemie, Berlin, 1960.
(6)   Pauling, L. C., "The Nature of the Chemical Bond and the Structure of Molecules and Crystals," 3rd ed., Cornell University Press, Ithaca, N. Y., 1960.
(7)   Sutton, L. E., "Tables of Interatomic Distances and Configuration in Molecules and Ions," Special Publication No. 18, p. M39s, The Chemical Society, London, 1965.
(8)   Palke, W. E., private communication.
(9)   Nelson, D. L., "Perspective Plotting of Two Dimensional Assays—PLOT 3D," University of Maryland, Department of Physics and Astronomy, College Park, Md.
(10) Coulson, C. A., *Nature* (1969) **221**, 1106.
(11) Marsmann, H., Van Wazer, J. R., Robert, J.-B., *J. Chem. Soc.* **1970**, 1566.
(12) Banyard, K. E., Hake, R. B., *J. Chem. Phys.* (1964) **41**, 3221.
(13) Boer, F. P., Lipscomb, W. N., *J. Chem. Phys.* (1969) **50**, 989.
(14) Keeton, M., Santry, D. P., *Chem. Phys. Lett.* (1970) **7**, 105.
(15) Moccia, R., *J. Chem. Phys.* (1964) **40**, 2186.
(16) Rothenberg, S., Young, R. H., Schaefer, H. F., *J. Amer. Chem. Soc.* (1970) **92**, 3243.
(17) Ruak, A., Csizmadia, I. G., *Can. J. Chem.* (1968) **46**, 1205.
(18) Pople, J. A., Santry, D. P., Segal, G. A., *J. Chem. Phys.* (1965) **43**, S129.
(19) Fogleman, W. W., Miller, D. J., Jonassen, H. B., Cusachs, L. C., *Inorg. Chem.* (1969) **8**, 1209.
(20) Mulliken, R. S., *J. Chem. Phys.* (1955) **23**, 1833.

(21) Wagman, D. D., Evans, W. H., Parker, V. B., Halow, I., Bailey, S. M., Schum, R. H., "Selected Values of Chemical Thermodynamic Properties," p. 14, National Bureau of Standards, Technical Note 270-3, U. S. Government Printing Office, Washington, D. C., 1968.

(22) Herzberg, G., "Molecular Structure and Molecular Spectra, III," p. 585, Van Nostrand, Princeton, N. J., 1966.

(23) Hollister, C., Sinanoglu, O., *J. Amer. Chem. Soc.* (1966) **88,** 13.

(24) Absar, I., Van Wazer, J. R., *J. Phys. Chem.* (1971) **75,** 1360.

(25) Unland, M. L., Letcher, J. H., Absar, I., Van Wazer, J. R., *J. Chem. Soc. (A),* **1971,** 1328.

(26) Marsmann, H., Groenweghe, L. C. D., Schaad, L. J., Van Wazer, J. R., *J. Amer. Chem. Soc.* (1970) **92,** 6107.

(27) Van Wazer, J. R., *Colloq. Int. Cent. Nat. Rech.* (1970) **182,** 27.

(28) Mulliken, R. S., *J. Chem. Phys.* (1962) **36,** 3428.

(29) Siegbahn, K., Nordling, C., Fahlman, A., Nordberg, R., Hamrin, K., Hedman, J., Johansson, G., Bergmark, T., Karlsson, S. E., Lindgren, I., Lindberg, B., "Atomic, Molecular and Solid State Structures Studied by Means of Electron Spectroscopy," pp. 94, 125, 128, and 132, Almquist and Wiksells Bok Tryckeri, Uppsala, 1967.

(30) Siegbahn, K., Nordling, C., Johansson, G., Hedman, J., Heden, P. F., Hamrin, K., Gelius, V., Bergmark, T., Werme, L. O., Manne, R., Baer, Y., "ESCA Applied to Free Molecules," p. 128, North Holland Publishing Co., Amsterdam, 1969.

(31) Robert, J.-B., Marsmann, H., Absar, I., Van Wazer, J. R., *J. Amer. Chem. Soc.* (1971) **93,** 3320.

(32) Edmiston, C., Ruedenberg, K., *Rev. Mod. Phys.* (1963) **35,** 457.

RECEIVED March 5, 1971.

3

Some New Ideas for Structural Organosulfur Chemistry

JEREMY I. MUSHER

Belfer Graduate School of Science, Yeshiva University, New York, N. Y.

The theory of hypervalent molecules is discussed with special reference to sulfur chemistry. A series of new organic sulfur(IV) compounds or "sulfuranes" and sulfur(VI) compounds or "persulfuranes" are predicted to be stable and are likely to possess interesting structural properties.

One of the functions of the theoretical chemist should be to stimulate his experimental colleagues by predicting new feasible chemical structures or new physicochemical phenomena. Nevertheless most of the real advances in chemistry are made by experimentalists, and the theoretician usually sits in the background quietly calculating his integrals. There are two reasons for this. The first is that chemistry is an experimental subject, and it is difficult for a scientist to develop the intuitive feel for problems of experimental significance without having spent time in the laboratory. In this the theoretician is at fault since instead of devoting the necessary effort to develop an understanding of the experiment—*i.e.*, the science itself—he hides himself behind his more and more inaccessible formalisms and his more and more intricate computer programs. The second reason for the theoretician's lack of influence on experimental chemistry is the skeptical attitude of the experimentalist whose general reaction to a new idea is "That's fine, now prove it!" In fact, it has not been considered legitimate to discuss ideas for experimental work without the explicit verification having already been carried out. All this has perhaps changed somewhat since Roald Hoffmann appeared on the scene. Almost single-handedly Hoffmann has shown the experimentalist that there is something to be learned from theory—although he was, of course, given respectability from his early collaboration with an experimentalist of some repute—and now, at least in special cases, it is permissible for a theoretician to publish experimental speculations. Hoffmann's subject, organic chemistry, has been greatly picked-over by theoreticians and experimentalists alike, so that new ideas are relatively rare, and the approach that Hoffmann and others must take requires great subtlety and imagination. Inorganic chemistry and metalloorganic chemistry have been rather neglected subjects so that there must still exist a number of wonderful ideas waiting to be developed. Ever since I stumbled upon the hypervalent molecules, I have felt that the recognition of their chemistry would prove eventually to be one of these wonderful ideas. Unfortunately, I have yet to "prove it" but these ideas have led to some amusing new iodine, xenon, and phosphorus chemistry (*1, 2, 3, 4*), and the potential is far from being tapped.

Some speculative structural organic chemistry of hypervalent sulfur —*i.e.*, the organic chemistry of tetravalent and hexavalent sulfur—will be discussed. While the structures I propose may look hypothetical, they are all exact analogs of known compounds already in the literature. There is no *a priori* reason why they all could not be prepared.

Hypervalent Molecules

The class of molecules which I refer to as hypervalent (*5, 6*) are those molecules which contain heteroatoms of Groups V-VIII of the Periodic Table in their higher valences. The argument for organic hypervalent molecules is simple. The first inorganic hypervalent molecules, PCl_5, $SeCl_4$, and ICl_3 were known in the early 1800's a full 100 years before the Lewis–Langmuir octet theory placed them virtually in intellectual oblivion. However the fact that organic molecules using these valences did not fit into any generally accepted theoretical scheme served to associate with such molecules the words "unusual" and "exceptional." Actually, there is nothing extraordinary about these valences, and when they are understood, it should be immediately possible to utilize them in organic chemistry.

I consider bonding in molecules to be best and most simply described in terms of bond orbitals which utilize only *p*-orbitals unless the number of bonds exceeds the number of *p*-electrons in the ground configuration, at which point hybridization with the *s*-orbitals must be introduced. There are two types of *p*-orbital bond:

(a) ordinary or covalent bonds in which a singly occupied *p*-orbital from the heteroatom is bonded to a singly occupied atomic orbital from the ligand

(b) hypervalent bonds in which a doubly occupied *p*-orbital from the heteroatom is bonded with *two* singly occupied atomic orbitals from two colinear ligands. The molecular orbital description of this three-center four-electron process involves doubly occupied bonding and non-bonding orbitals which are written equally well in terms of localized equivalent orbitals.

The essential difference between these bonds is that a single hetero-atom orbital is used to form two hypervalent bonds in complete disregard of traditional valence bond ideas. This permits hybridization to be intro-duced only when it actually is needed, as in reaching the highest valences such as the valence 6 of sulfur. The right-pyramidal structure of SF_4 is explained by this theory by considering atomic sulfur to have the Hund's rule configuration $s^2 p_x^2 p_y p_z$. The singly occupied p_y- and p_z-orbitals are used to form the two equatorial bonds, and the doubly-occupied p_x-orbital is used to form the two colinear axial bonds. Notice that I have not mentioned d-orbitals. In this I take the pragmatic position that I see no need to introduce them, and, there is really nothing that d-orbitals can or cannot do that would modify the qualitative conclusions of the theory. (Meaningful quantitative results can never be obtained from any such simple-minded theory with or without d-orbitals.) Thus rather than either put up the straw man or knock him down once he is up, I instead observe that he is, after all, only made of straw.

The essential criterion for stability of hypervalent bonds is the extent to which the interactions of the three atomic orbitals lower the energy of the bonding orbital. Sufficient energy lowering has been shown (7, 8) to occur when the ligands have high electron affinities and the heteroatoms have low ionization potentials. When sulfur is the heteroatom, it appears that not only the atoms F and Cl are sufficiently electronegative to form hypervalent bonds but also it is, or should be, possible to form such molecules with OR, C_6H_5, CF_3, $CH{=}CH_2$, $C{\equiv}CH$, NR_2, and other ligands. In short, except for unstrained saturated carbon it appears that most common functional groups can form stable hypervalent bonds with sulfur. Such bonds will be weak, however, so that the molecules will often be thermodynamically unstable to unimolecular or bimolecular de-composition. If they are flexible enough to allow easy kinetic pathways, they will not be easily observed or studied, but that, too, is one of their interesting features. To stabilize these molecules it is advantageous to cyclize them, and most of the examples discussed here will be of such stabilized cyclic systems.

Sulfuranes and Persulfuranes

Consider derivatives of the two hypothetical molecules which I call sulfurane (1) and persulfurane (2). (A new name would be required to describe ylides of the form $R_4S{=}CR'_2$. By analogy these could be called persulfoniumylides.) The theory of hypervalent molecules argues that a sulfurane will be stabilized when its colinear or hypervalent bonds are directed towards electronegative atoms, and the persulfuranes will be stabilized when several if not all of its bonds are directed towards elec-

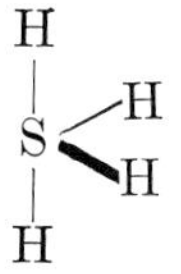

$$(1) \qquad\qquad (2)$$

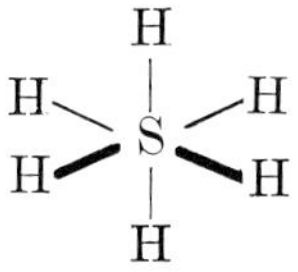

tronegative atoms. Few simple sulfuranes are known—*e.g.*, perfluoro-sulfurane (sulfur tetrafluoride), perchlorosulfurane (sulfur tetrachloride), dimethylaminotrifluorosulfurane (9) which is one of the two isomers (3a) or (3b), and bistrifluoromethyldifluorosulfurane (10) which is one

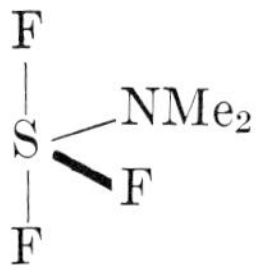

$$(3a) \qquad\qquad (3b)$$

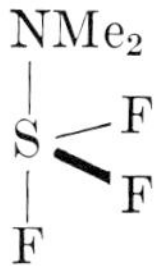

of the four isomers (4a,b,c,d) or a mixture thereof. A larger number of persulfuranes have been made in addition to the well-known perfluoro-

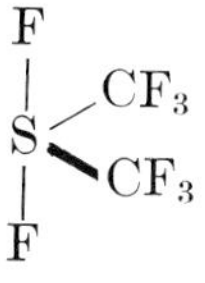

$$(4a) \qquad\quad (4b) \qquad\quad (4c) \qquad\quad (4d)$$

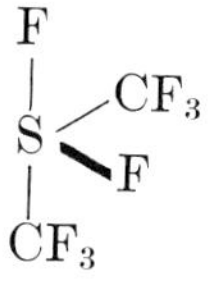

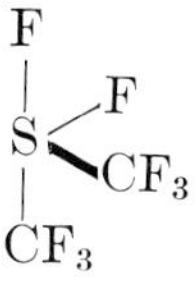

persulfurane (sulfur hexafluoride). These are generally of the form RSF_5 with $R=OSF_5$, COF, OCOF, CCH, $CH_2CH(OCH_3)_2$, and CH_2CH-$(OCOCH_3)_2$ (11, 12). There have, however been no amino compounds, no cyanides, isocyanides, organic esters, or methyl derivatives and few more complex molecules.

It is possible that unpublished attempts to isolate some more complex sulfuranes have failed as a result of their instability to intramolecular rearrangement or to hydrolysis—*e.g.* as in the following equation—but at least some of these have been postulated as reactive intermediates (13), and it is not unreasonable to assume that suitable conditions will be found under which they can be isolated.

$$
\begin{array}{l}
\overset{\displaystyle OCH_3}{\underset{\displaystyle F}{\overset{\displaystyle |}{\underset{\displaystyle |}{S}}}}\!\!<\!\!\begin{array}{l} C_6H_5 \\ C_6H_5 \end{array} \quad \rightarrow (C_6H_5)_2SO + CH_3F
\end{array}
$$

$$
\begin{array}{l}
\overset{\displaystyle OCOCH_3}{\underset{\displaystyle OCOCH_3}{\overset{\displaystyle |}{\underset{\displaystyle |}{S}}}}\!\!<\!\!\begin{array}{l} CH_3 \\ CH_3 \end{array} \quad \underset{H_2O}{\rightarrow} \quad (CH_3)_2SO + CH_3COOH
\end{array}
$$

Consider now some cyclic sulfuranes using sulfur–carbon bonds. A simple set of isomeric molecules would be the three trimethylenedifluorosulfuranes (5a,b,c) for which systematic names could be given once they

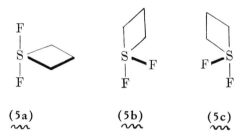

(5a)   (5b)   (5c)

were prepared. The first could be prepared simply by fluorinating trimethylene sulfide, but the others presumably would have to be obtained *via* cyclization with SF$_4$. The following observations should be made:

(1) It is not known *a priori* which of the structures is energetically preferable;

(2) It is not known *a priori* whether the barrier to intramolecular inversion is sufficiently low to allow equilibration among them.

The first observation is in contrast to the studies on phosphoranes (*14, 15, 16*) for which the equatorial cycle is highly unfavorable owing to the non-cyclic equatorial C–P–C angle of 120°. The non-cyclic C–S–C bond angle is presumably in the vicinity of the 102° F–S–F angle of SF$_4$. [For relevant data *see* Table VI of Reference (*17*).] Also trimethylenesulfide is a known compound. This is an example of the difference between the pure *p*-bond picture of sulfuranes and the *sp*-hybridized picture of phosphoranes. The only evidence relative to the second observation is the lack of pseudorotation in solvolysis sulfurane intermediates (*16*) although some authors seem intuitively to believe pseudorotation likely in these compounds (*18*).

Spiro systems should have even greater stability, and examples of these are the [3.3]-thiaspiraranes **(6a,b)** and the [4.4]-thiaspirarenes **(7)** of the same two geometries. [For the name, spirarene, *see* References (*19*) and (*20*).] These could in turn be chlorinated to give the *S,S-cis-*

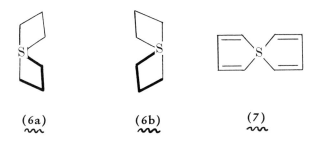

(6a)   **(6b)**   **(7)**

dichloro-[3.3]-thiaspirarane **(8a)** and the *S,S-cis-*dichloro-[4.4]-thiaspi-rarene **(9a)** and their optical isomers while the trans isomers **(8b)** and **(9b)** presumably would require different synthetic procedures. The *cis*-persulfuranes then can be cyclized further to give the *d,l*-pairs of [3.3.3]-thiaspirarane (tristrimethylenesulfur) and [4.4.4]-thiaspirarene.

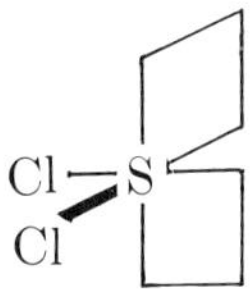

(8a) 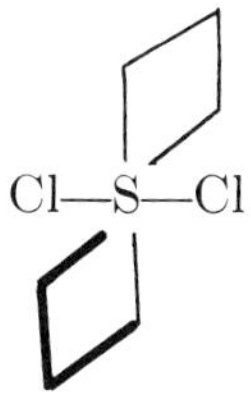   **(8b)** 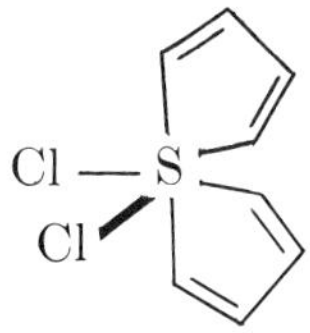   **(9a)** 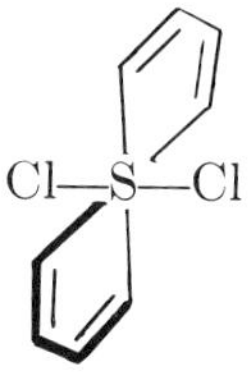   **(9b)**

Consider now some cyclic sulfuranes with oxygen–sulfur bonds similar to the several thiaoxaspiraranes discussed previously (*5*, *6*). The

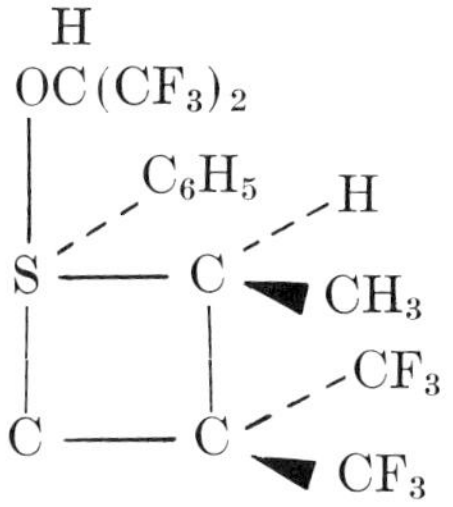

(10) 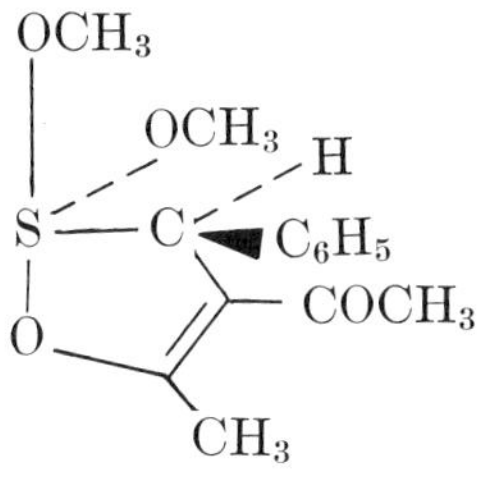   **(11)**

oxysulfuranes (10) and (11) and their diastereo and configurational isomers would be particularly amusing to prepare as they are exact analogs of the oxyphosphoranes (21, 22, 23) prepared by Ramirez and Westheimer. These authors' studies on conformational stability could be repeated for the sulfuranes although again the barrier to intramolecular inversion is expected to be high. An amusing cyclic oxysulfurane would be the analog of the iodosodilactone (12) (1) of structure (13). The successful synthesis of (13) would bring us a step closer to the preparation of (14) which I have long considered a prime candidate for a penta-coordinate nitrogen compound (5, 6).

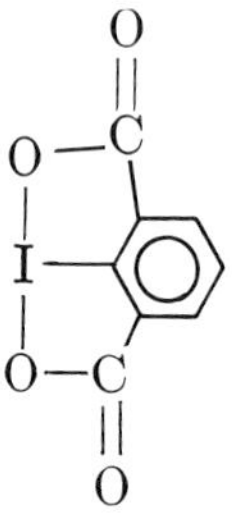

(12) 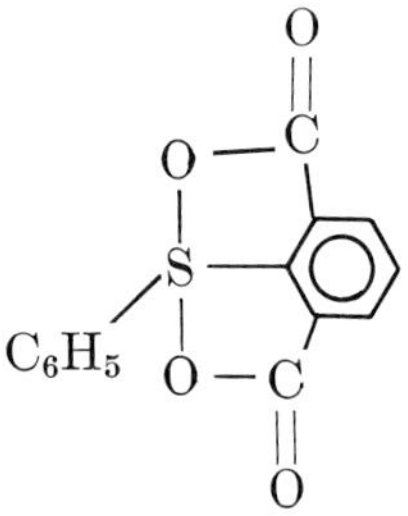   (13) 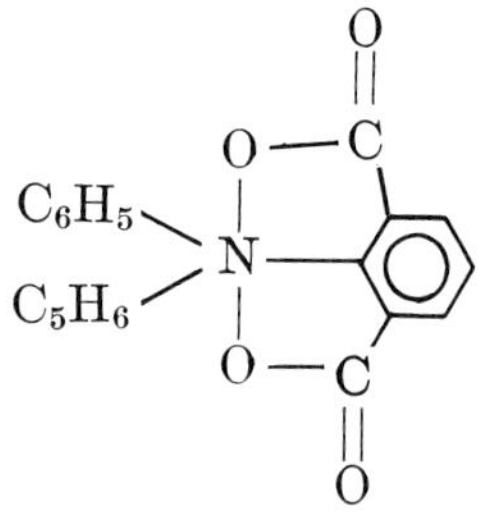   (14)

As a last example of cyclic sulfuranes, consider the analogs to the 1,3-diazafluorophosphetidines (15) prepared by Schmutzler (24) and MacDiarmid (25) according to the reaction

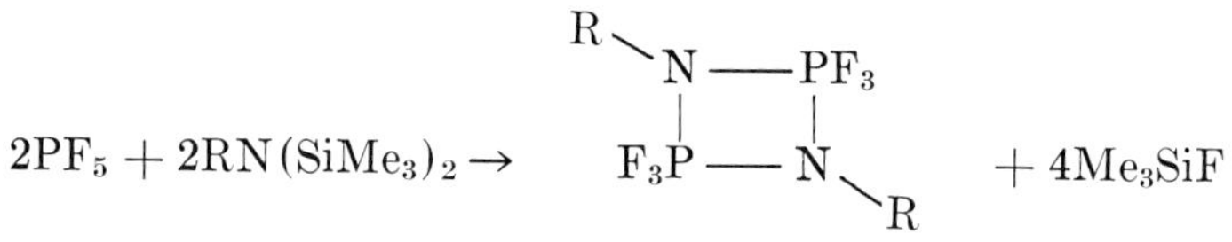

(15)

It is not unreasonable to expect that the reaction of SF_4 with $RN(SiMe_3)_2$ would give the three fluoro-1,3-dithiadiazetidines (16a,b,c) where for simplicity only, it is assumed that there is rapid inversion about the nitrogen atom. Notice that there is only one diazaphosphetidine (15) since the four-membered ring cannot be diequatorial, and since the equatorial–axial ring provides a plane of symmetry, again pointing to the difference between the pure p-bonding of the sulfuranes and the sp-hybridized bonding of the phosphoranes.

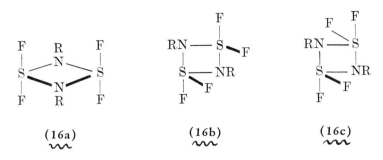

(16a)   (16b)   (16c)

Since this article was completed, several papers on sulfuranes have appeared in the literature, and the term "sulfurane" has been proposed independently and used for the first time (26). Martin and Arhart (26) have prepared $(C_6H_5)_2S[OC(CF_3)_2C_6H_5]_2$ and are in the process of determining its x-ray structure. Sheppard (27) has prepared perfluoro-tetraphenyl sulfurane, and Kapovits and Kálmán (28) have prepared a spirodiacycloxysulfurane of the structure (12) proposed by the author in References 5 and 6 and are in the process of determining its x-ray structure. Also La Rochelle and Trost (29) have presented an extended study showing the intermediacy of sulfuranes in organic synthesis, and this article gives a survey of the previous literature on the problem. The author has prepared a sequel to the present article which deals with the hypervalent chemistry of selenium and tellurium (30).

Literature Cited

(1) Agosta, W. C., *Tetrahedron Lett.* (1965) 2681.
(2) Livingston, H. K., Sullivan, J. W., Musher, J. I., *J. Polymer Sci.* (1968) C22, 195.
(3) Musher, J. I., *J. Amer. Chem. Soc.* (1968) 90, 7371.
(4) Musher, J. I., "Conformational Analysis," p. 177, G. Chiurdoglu, Ed., Academic Press, New York (1971).
(5) Musher, J. I., *Angew. Chem.* (1969) 81, 68.
(6) Musher, J. I., *Angew. Chem. Int. Ed.* (1969) 8, 54.
(7) Pitzer, K. S., *Science* (1963) 139, 414.
(8) Musher, J. I., *Science* (1963) 141, 736.
(9) Demitras, G. C., MacDiarmid, A. G., *Inorg. Chem.* (1967) 6, 1903.
(10) Sauer, D. T., Shreeve, J. M., *Chem. Commun.* (1970) 1679.
(11) Czerepinski, R., Cady, G. H., *J. Amer. Chem. Soc.* (1968) 90, 3954.
(12) Hoover, F. W., Coffman, D. D., *J. Org. Chem.* (1964) 29, 3567.
(13) Corey, E. J., Durst, T., *J. Amer. Chem. Soc.* (1968) 90, 5548, 5553.
(14) Westheimer, F. H., *Accounts Chem. Res.* (1968) 1, 70.
(15) Ramirez, F., *Accounts Chem. Res.* (1968) 1, 168.
(16) Mislow, K., *Accounts Chem. Res.* (1970) 3, 321.
(17) Baenziger, N. C., Buckles, R. E., Maner, R. J., Simpson, T. D., *J. Amer. Chem. Soc.* (1969) 91, 5749.
(18) Muetterties, E. L., Phillips, W. D., *J. Chem. Phys.* (1967) 46, 2861.
(19) Ashe III, A. J., *Tetrahedron Lett.* (1968) 359.
(20) Hoffmann, R., *Accounts Chem. Res.* (1971) 4, 1.

(21) Ramirez, F., Smith, C. P., Pilot, J. F., *J. Amer. Chem. Soc.* (1968) **90,** 6726.
(22) Gorenstein, D., Westheimer, F. H., *Proc. Natl. Acad. Sci. U.S.* (1967) **58,** 1747.
(23) Ramirez, F., Pilot, J. F., Madan, O. P., Smith, C. P., *J. Amer. Chem. Soc.* (1968) **90,** 1275.
(24) Schmutzler, R., *Chem. Commun.* (1965) 19.
(25) Demitras, G. C., Kent, R. A., MacDiarmid, A. G., *Chem. Ind. (London)* (1964) 1712.
(26) Martin, J. C., Arhart, R. J., *J. Amer. Chem. Soc.* (1971) **93,** 2341.
(27) Sheppard, W. A., *J. Amer. Chem. Soc.* (1971) **93,** 5597.
(28) Kapovits, I., Kálmán, A., *Chem. Commun.* (1971) 649.
(29) LaRochelle, R. W., Trost, B. M., *J. Amer. Chem. Soc.* (1971) **93,** 6077.
(30) Musher, J. I., *Ann. N. Y. Acad. Sci.,* in press, 1971.

RECEIVED March 5, 1971. Supported in part by the National Science Foundation.

4

The Spectrum of Sulfur and its Allotropes

B. MEYER, M. GOUTERMAN, D. JENSEN, T. V. OOMMEN, K. SPITZER, and T. STROYER-HANSEN

University of Washington, Seattle, Wash. 98105

Orthorhombic sulfur absorbs at 285 and 265 nm because of S_8; polymeric sulfur absorbs at 360 nm; liquid sulfur has an absorption edge which shifts from 400 nm at 120°C to 700 nm at 700°C. The shift is a result of changes in molecular composition and simultaneous absorption by cyclo-S_8, polymeric sulfur, S_4, and S_3. The spectrum of sulfur vapor changes with temperature and pressure. Blue absorption is caused by S_2, green absorption by S_8, absorption at 400 nm by S_3, and absorption at 520 nm by S_4. The comparison of spectra of allotropes in various phases, including solutions and matrices, with extended Hückel calculations shows that the spectrum of chains converges towards the near infrared with increasing molecular size, while sulfur rings all absorb in the near ultraviolet.

Sulfur exists in a variety of metastable molecular forms and polymorphs (1). For most of these allotropes the color is known well and is used often as an identifying tool. However, the spectrum of these species, with few exceptions, is not established well. This article reviews the spectra observed for polymeric sulfur, S_8, S_6, S_4, S_3, and S_2 and discusses and predicts the electronic energy levels, the spectrum, and color of these and some other allotropes.

This paper consists of five sections. In the first, the absorption spectrum of solid sulfur is discussed; the second deals with liquid sulfur, the third reviews spectra of the vapor, and the fourth describes spectra of sulfur allotropes in solution, glasses, and matrices. The fifth section is a short review of some extended Hückel computations of energy levels of various allotropes in different structural configurations.

Solid Sulfur

Orthorhombic S_8, the stable form of sulfur at room temperature, forms yellow crystals which are transparent above 400 nm. In thick crystals, absorption occurs above 400 nm. The spectrum has been recorded by Clark and Simpson (2) by reflection and absorption spectroscopy. Figure 1 shows the absorption spectrum, with a maximum at 285 nm which is explained as being a result of the first allowed transition of cyclo-S_8. The energy levels of S_8 are discussed later. The first absorption peak at 285 nm is so strong that crystals are opaque below 350 nm. Thus, the known ultraviolet spectrum of sulfur consists of an absorption edge. The vacuum ultraviolet spectrum of solid sulfur is not known. Fukuda (3) studied the ultraviolet absorption of sulfur and discovered that the absorption edge shifts with temperature at a rate of 0.2 nm/$^\circ$K between 0°C and the melting point. Our observations (4) indicate that the shift between 77°K and 25°C is 0.23 nm/$^\circ$K. This shift explains why the color of sulfur fades reversibly upon cooling. At 77°K orthorhombic sulfur is snow white like table salt (5).

The shift of the absorption edge and, thus, the color change is caused by change in thermal population of the vibrational levels of the electronic ground state of S_8 (3). A quantitative correlation between the shift of the absorption edge and kT has been developed by Gilleo (6) for selenium. This model has been applied to sulfur by Bass (7).

Knowledge of electronic energy levels of sulfur allotropes is still rather meager. It is amazing that no great effort has been made to record and analyze the spectrum of S_8. For example, S_8 must have a low lying triplet state in the visible. This triplet state is not known although its

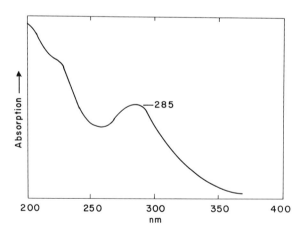

Figure 1. Absorption spectrum of a thin film of S_8
at 25°C (2)

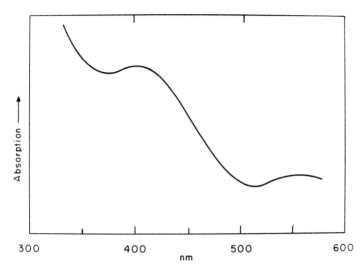

Figure 2.   Absorption spectrum of red sulfur glass at 77°K, obtained by quenching a boiling film of sulfur in liquid nitrogen

transition energy and lifetime are important keys to understanding the photolysis of the S–S bond. S–S scission occurs in the presence of daylight and is responsible for many perplexing properties of S_8 in chemistry, radiation biology, and physiology. Thus, the discovery of triplet state properties of S_8 should be one of our foremost goals.

The spectra of the unstable solid allotropes of sulfur are not known well. S_6, S_7, S_8, S_{12}, and other species (8, 9, 10) form yellow to white solids. Their solution spectra are discussed later. Quenched liquid sulfur or vapor produces yellow polymeric sulfur. Its spectrum shows a shoulder at 360 nm, as shown later. If liquid sulfur or sulfur vapor is quenched rapidly to 77°K or below, deep colored films are formed. The spectrum of red sulfur produced by quenching a boiling liquid film is shown in Figure 2. The color is a result of three spectral features: the absorption edge of polymeric sulfur, an absorption peak at 400 nm which is caused by the lowest allowed electronic transition of S_3, and an absorption peak at 550 nm which is a result of S_4. The spectrum of trapped vapor consists of several broad peaks; it changes with the composition of the vapor source and depends on speed of deposition and many other factors. The deep color of all trapped metastable species fades at −90°C to deep yellow. Above this temperature, the spectrum shows only polymeric sulfur and S_8.

The study of the spectrum of solid sulfur is not only complicated by metastable allotropes but also by impurities. Even 99.999% sulfur contains traces of organic matter which upon melting react with sulfur and color it brown. This brown color has been mistaken frequently as stemming from metastable allotropes. However, the formation of brown polysulfides is irreversible and can be distinguished easily from unstable allotropes which at room temperature all convert into yellow orthorhombic sulfur.

Freshly poured vats of Frasch sulfur have a blue hue (11) which disappears within a few days. We believe that this color is partly a result of the reflection spectrum of S_3 and S_4. Although liquid Frasch sulfur can contain only minute traces of these species, their concentration is sufficient to explain this hue. The slow speed of fading is caused by the slow kinetics of recombination of S_3 and S_4 trapped in a matrix of S_8; it does not indicate that S_3 or S_4 is stable at room temperature.

Liquid Sulfur

At the melting point, liquid sulfur is transparent and has the same yellow color as the crystals. At the boiling point it is deep red and opaque. The absorption coefficient in the ultraviolet is so high that the spectrum of even a very thin film consists only of an absorption edge. Figure 3 shows the absorption edge of a 160-μ thick film at various temperatures. The spectra above 444°C, the normal boiling point, were taken at high pressure in sealed quartz cells. Figure 4 shows the shift rate as a function

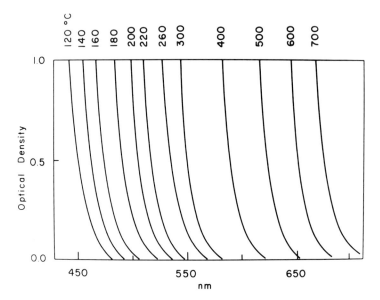

Figure 3.   Red absorption edge of liquid sulfur at various temperatures between 120°–700°C

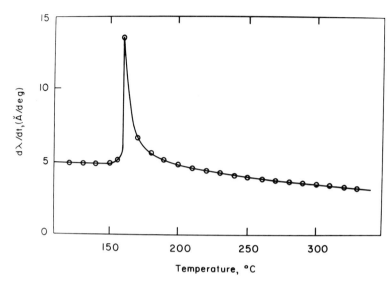

Figure 4. Shift rate, in per degree K, of absorption edge of liquid sulfur at optical density 0.6

of temperature (*4*). Of the earlier workers, Mondain-Monval (*12*) observed a shift of 0.294 nm/°K below 160°C and 0.74 nm/°K above 160°C; Bass (*7*) found an average value of 0.62 nm/°K between the melting point and 300°C compared with our value of 0.576 nm/°K. Bass and earlier workers explained the shifts in terms of thermal population of ground state vibrational modes. Such population explains the temperature dependence of the spectrum of solid sulfur. In the case of liquid, however, the shift rate is more than twice that of the solid, and Figure 4 shows that the shift is not smooth. The discontinuity at 160°C reminds one of the extreme viscosity change at that temperature (*13*). This viscosity change is caused by polymerization of S_8 and change in molecular composition. We have restudied recently the spectrum of liquid sulfur and concluded that at the melting point the color is mainly a result of thermally broadened S_8. At 160°C, it is caused by overlap of S_8 and polymer absorption. At the boiling point the color is caused by at least four species: the absorption of thermally broadened S_8, thermally broadened polymeric sulfur, an absorption peak caused by an electronic transition of S_3 at 400 nm, and absorption by S_4 at 550 nm. The assignment of the visible peaks to S_3 and S_4 is based on comparison of spectra of the liquid with those of sulfur vapor, sulfur solutions, as well as glasses and matrices containing selected sulfur allotropes (*14*). From these we conclude that liquid sulfur contains 0.1 to 3% S_3 and S_4. Since neither the composition of liquid sulfur nor the spectra of all potentially important allotropes is known fully, it is possible that we have overlooked other less important sources for the color, such as S_6, S_7, or S_{10}.

The discovery of shoulders and peaks in the visible spectrum of hot liquid sulfur (Figure 5) and the assignment to S_3 and S_4 have shown how little research has been done and how little is known about the composition of liquid sulfur. There is no question that this phase contains a wealth of unearthed treasures for physicists, physical chemists, as well as synthetic chemists of all types.

The spectrum yields some important insights into the type of species the liquid contains and into their reactivity. The fifth section shows that the red absorption indicates that liquid sulfur contains reactive chains because sulfur rings in analogy with alkanes are expected to be yellow and to absorb in the near ultraviolet, while sulfur chains in analogy with alkyl diradicals are expected to be colored deeply.

Sulfur Vapor

Below 250°C the spectrum of saturated vapor has been described by Bass (7). It consists of unresolved maxima at 210, 265, and 285 nm which are caused by electronic transitions of cyclo-S_8. If traces of impurities are present, a long vibrational progression appears between 185–210 nm with an origin at 45950 cm^{-1} as a result of CS_2 which is formed in the reaction between any hydrocarbon and sulfur. This spectrum is an even

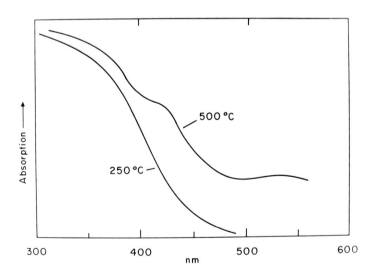

Figure 5.   Absorption spectrum of liquid sulfur at 250° and at 500°C

Table I. Electronic Energy Levels of S_2 (*34*)

State		Electron Configuration $\sigma\ \pi\ \pi\ \sigma$	Transition Energy observed (cm^{-1})	estimated	$\Delta G_{1/2}$ (cm^{-1})
c'	$^1\Sigma_u{}^+,{}^1\Delta_u$	2 4 1 0	$55{,}448.3 + (?)x$	$(?)\sim59{,}900$	785
			$(?)y$	$(?)\sim64{,}000$	
c	$^1\Sigma_u{}^+$	2 4 1 0	$51{,}401.3 + (?)y$	$(?)\sim59{,}900$	814
D	$^3\Pi_{g,r}$	(?)2 4 1 0	58,750	58,750	786
e	$^3\Sigma_u{}^-$	2 4 1 0	56,983.6	56,984	—
d	$^1\Delta_u$	2 4 1 0	$52{,}244.7 + x$	$\sim56{,}700$	811
C	$^3\Sigma_u{}^-$	2 4 1 0	55,633.3	55,633	822
b	$^1\Delta_u$	2 3 3 0	$36{,}743 + x$	$\sim41{,}200$	432.7
b'	$^1\Pi_g$	1 4 3 0	$13{,}451.8 + a$	$\sim37{,}000$	533.7
B'	$^3\Pi_{g,i}$	1 4 3 0	$14{,}144.7 + A'$	$\sim36{,}000$	—
B	$^3\Sigma_u{}^-$	2 3 3 0	31,689	31,689	428.5
B''	$^3\Pi_u$	(?)2 4 1 1	$\leqslant31{,}700$	$\leqslant31{,}700$	—
A	$^3\Sigma_u{}^+$	2 3 3 0	$697 + A'$	$\sim22{,}550$	477
a	$^1\Sigma_u{}^-$	2 3 3 0	a	$\sim23{,}550$	—
A'	$^3\Delta_{u,i}$	2 3 3 0	A'	$\sim21{,}855$	483.4
y	$^1\Sigma_g{}^+$	2 4 2 0	y	$\sim8{,}500$	(?)693
x	$^1\Delta_g$	2 4 2 0	x	$\sim4{,}500$	695.91
X	$^3\Sigma_g{}^-$	2 4 2 0	0	0	719.98

more sensitive indication of impurities than the infrared spectrum of CS_2 (*15*).

At higher temperature and at pressures below 1 torr, the vapor spectrum is characterized by the Schumann–Runge bands of S_2 which, with increasing temperature, extend into and throughout the visible and give a violet hue to the vapor. The origin of these $^3\Sigma_u{}^- \leftarrow {}^3\Sigma_g{}^-$ bands is at 31835 cm^{-1}. Emission spectra of S_2 vapor show other transitions, among them the transition $^1\Delta_u \rightarrow {}^1\Delta_g$ which originates at 36743 cm^{-1}. The spectrum of S_2 has been reviewed well by Barrow (*16*). The properties of the states of S_2 are summarized in Table I. Emission spectra contain also atomic lines which result from transitions between the energy levels indicated in Table II.

At intermediate pressures and temperatures, sulfur changes its color. Saturated vapor between 350° and 500°C is deep red. At 1–10 torr and 500°–650°C, it is yellow-orange. This is caused by an absorption system between 360–440 nm consisting of progressions of vibrational bands and a continuum between 480–600 nm.

We show later that the spectrum around 400 nm is a result of S_3 and that at 520 nm it is caused by S_4. The temperature dependence of the intensity of these systems is shown in Figure 6.

The vibrational bands of the 400 nm system were described first by Graham (17), and assigned to S_3 or S_4 by d'Or (18). The assignment was based on correlation of intensity with vapor pressure measurements made by Preuner (19). Recently the vapor composition has been confirmed mass spectroscopically (20, 21), and the spectrum was restudied carefully (22). We have studied this system with pure $^{34}S_x$ and $^{32}S_x$ vapor and determined that the origin is at 23465 cm^{-1}. A tracing of a photograph of this system is shown in Figure 7. A vibrational analysis is given in Table III. It yields $\nu' = 420$ cm^{-1} and $\nu'' = 590$ cm^{-1}, indicating that only one of the stretching modes is active in this transition. A rotational anlysis is difficult because the B values are small, and the band heads are overlapped. The electronic character of this transition is discussed later. The continuous absorption system at 520 nm was assigned first to S_4 by Braune (23) who measured the pressure of unsaturated vapor. Assignment of the spectrum had to wait for the discovery of a synthetic method to make pure S_3 and S_4 because the vapor always contains both in comparable concentrations. As discussed below, we have recently (24) suc-

Table II.   Energy Levels of Sulfur Atom (35)

		Level	
Designation	J	*cm^{-1}*	*ev*
$^3P(3p^4)$	2	0.0	0
	1	396.8	0.049
	0	573.6	0.071
$^1D(3p^4)$	2	9239.0	1.15
$^1S(3p^4)$	0	22181.4	6.5
$^5S°(4s)$	2	52623.88	6.5
$^3S°(4s)$	1	55331.15	6.9
$^5P(4p)$	1	63446.36	
	2	63457.33	~7.8
	3	63475.26	
$^3P(4p)$	0	64891.71	
	1	64889.23	~8.0
	2	64892.89	
$^3D°(4s')$	1	67816.87	
	2	67825.72	~8.3
	3	67843.38	
$^5D°(3d)$	4	67878.03	
	3	67890.45	
	2	67888.25	~8.4
	1	67885.97	
	0	67884.67	

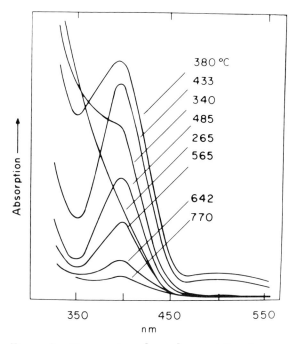

Figure 6. Temperature dependence of the absorption spectrum of sulfur vapor at 15 torr between 300–550 nm. The spectra were recorded so that the vibrational structure of the 400 nm band is suppressed and only the envelope of this system appears.

ceeded in making pure S_3 and pure S_4 in matrices and frozen glasses and have shown that the vibrational spacing and electronic origin of the 400 nm gas phase system is identical with that of S_3 in matrices. Likewise, the continuum around 520 nm is caused by S_4. Comparison with solid quenched sulfur and hot liquid sulfur has shown that both contain S_3 and S_4 (*14*).

Solutions, Frozen Solutions, and Matrices

Only S_8, S_6, and S_2 can be made in reasonably stable pure form. All other allotropes exist only in solution or as components of complex mixtures. This section discusses the spectra of solutions containing pure allotropes.

The spectrum of a $10^{-3}M$ solution of S_8 in isopentane–methylcyclohexane (iPC) at 77°K and 25°C is shown in Figure 8. The same figure shows also the spectrum of S_6. Because of the red tail of the absorption edge, S_6 forms darker crystals than S_8. The spectrum of polymeric sulfur in glycerol is shown in Figure 9. It shows a shoulder at 360 nm which is not present in S_8 or S_6. The spectrum of cyclo-S_7, S_{10}, S_{12}, and other recently prepared or isolated metastable rings (*10*) is not known yet, but the solutions are pale yellow or light green. Species with less than six atoms are not stable in ordinary solution but must be trapped in a frozen solution or matrix.

The spectrum of S_2 is obtained in rare gas matrices, in methane, and in SF_6 at $20°K$, by deposition of a molecular beam of S_2 or by *in situ* photolysis of S_2Cl_2 (*25*). S_2 is not stable in an organic glass at $77°K$. In contrast, S_3 is stable in an isopentane–methylcyclohexane glass at $77°K$ and is prepared easily by photolysis of S_3Cl_2. Figure 10a shows the spectrum of a $10^{-3}M$ solution of S_3Cl_2 in iPC. Figure 10b is the spectrum of the same solution after 5 minutes of photolysis with a 2 kw high pressure mercury arc. Figure 10c shows the photolysis product in Kr at $20°K$. The origin of the transition is at 430 nm and the vibrational frequency is $\nu'_1 = 420$ cm^{-1}. Both values correspond to that of the green vapor system at 400 nm and the absorption of frozen liquid sulfur. Figures 10d,e show that S_4Cl_2 undergoes analogous photolysis. The continuous absorption of the photolysis products matches that of sulfur vapor. S_5 and longer chains have not been prepared yet. It is feasible to use photolysis of chlorosulfanes to prepare S_5 and larger molecules, but we have noted that S_3 and S_4 are susceptible to secondary photolysis.

Comparison with Computation on Energy Levels of Allotropes

Theoretical studies of sulfur compounds have been done by Clark (*2*), Cusachs (*26*), Feher (*27*), Buttet (*28*), and Palma (*29*). These

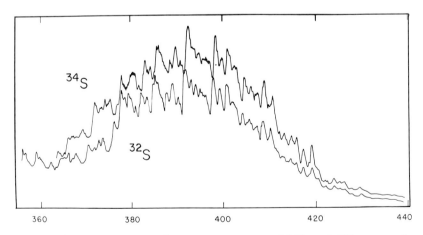

Figure 7. Tracing of an absorption spectrum of $^{34}S_3$ and $^{32}S_3$ vapor absorption photographed on a 0.75 m Czerny-Turner spectograph

Table III. Vibrational Assignment for S_3

(Position of Band Heads in cm^{-1})

v''_1 / v'_1	0	1	2	3	4
0	23465	—	—	—	—
1	23875	23281	—	—	—
2	—	—	—	—	—
3	24736	24123	—	—	—
4	25128	24575	23988	—	—
5	25519	24947	24401	—	—
6	25943	25363	—	—	—
7	—	25784	—	24678	24044
8	26750	26152	25599	25044	24497
9	—	26579	25999	—	24878
10	—	—	26431	—	25295
11	—	—	26847	26245	25698
12	—	—	—	—	26074

studies have considered various problems of individual sulfur allotropes. Since we have been attempting to understand a range of spectral observations involving many different allotropes and cases where the allotrope often is characterized poorly, it seemed useful to calculate the orbital energy diagram for a large number of possible allotropes using one simple model. For this type of exploratory study, the extended Hückel model seemed suited ideally.

The extended Hückel calculations follow the same mathematical form used earlier for the study of metalloporphyrins (30). The actual program used was a totally revised version of the one used at that time (31). The extended Hückel parameters for sulfur and hydrogen are given in Table IV. We did not include the empty $3d$ orbitals. The form of the Hamiltonian is

$$H_{pq} = \tfrac{1}{2}(H_{pp} + H_{qq})S_{pq}[\kappa + (1 - \kappa)\delta_{pq}]$$

where S_{pq} is the overlap. Calculations were done with $\kappa = 1.89$. We made no attempt to optimize κ because of the exploratory nature of the calculations. The values of H_{pp} were adjusted to suit the net atomic charge by an iterative self-consistent charge procedure in the manner described earlier (30). For the cyclic compounds S_6, S_7, S_8, and S_{12}, the calculations used the x-ray geometries. In S_2 the bond distance is 0.1889 nm; in S_3 it is 0.198 nm, and the S_3 bond angle is 120°. Longer diradical chains are assumed to be planar with a bond distance of 0.204 nm and a bond angle of 104°. Sulfanes have an H–S distance of 0.140 nm, an H–S angle of 95°, an S–S distance of 0.204 nm (except for H_2S_2 where it is 0.195 nm), and an S–S angle of 104°. The various forms of S_4 are based on the following data. In the planar ring the S–S distance is 0.206 nm, in the planar branched structure it is 0.195 nm, and in all others it is 0.20 nm. The S–S angle is 90° in the ring, 120° in the branched structure, and in all others 112°.

Before we present the results of these calculations, it is useful to consider the expected results. Sulfur rings are isovalent to cycloalkanes $(CH_2)_n$ and might be expected to have large gaps between the top filled and lowest empty orbitals. The sulfanes HS_nH, formerly called poly-sulfides, would be expected also to have large gaps between the top filled and lowest empty orbitals in analogy to the isovalent linear alkanes. On the other hand, the open chain S_n allotropes are isovalent to alkane di-radicals and would be expected to be colored. We considered also the possibility of branched chain allotropes, whose properties we could not predict in advance. The extended Hückel calculation was used to see whether the expected properties were supported by the simplest model orbital calculation, to determine the dependence on number of sulfur atoms, and to see if branched chain structures are reasonable. Moreover,

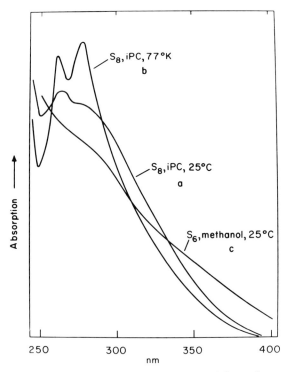

Figure 8.   Absorption spectrum of $10^{-3}M$ solution containing (a) S_8 in isopentane–methylcyclohexane (iPC) at 25°C; (b) S_8 in iPC at 77°K, and (c) S_6 in methanol at 25°C

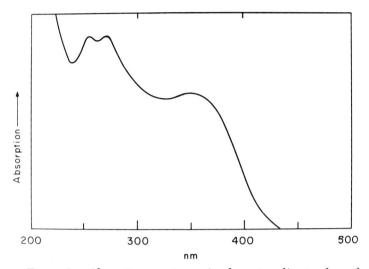

Figure 9. Absorption spectrum of polymeric sulfur in glycerol at 25°C

polymeric sulfur exists, and it was hoped that extrapolation from the results on finite S_n systems would allow its spectrum to be explained.

The extended Hückel results answer these questions. The sulfanes form a regular series. As shown in Figure 11d, the calculations predict colorless compounds whose absorption edge converges toward 400 nm with increasing chain length. The experimental data confirm this prediction. The cyclic alkanes (Figure 11a) are calculated to show a large energy gap between top filled and lowest empty orbitals (32). However, there is no regular shift of absorption edge with ring size, and of the computed molecules S_6 is yellow-pink while all others are yellow.

The sulfur chains are different from cyclic rings or sulfanes. As with O_2, the top filled orbitals of S_2 are theoretically group degenerate, and the ground state is a triplet. The lowest allowed transition is from a more deeply buried orbital to the lowest partly filled orbitals. In S_3 there are two orbitals separated by 0.8 ev available for the two highest energy electrons of the ground state. The ground state is a triplet with both

Table IV. Extended Hückel Sulfur Parameters

	α_s	α_p	ζ_s	ζ_p
S^+	−33.8	−23.4	2.1223	1.8273
S°	−20.3	−10.4	2.1223	1.8273
S^-	−11.1	−2.1	2.1223	1.8273
H^+	−15.3568	—	1.00	—
H°	−13.5950	—	1.00	—
H^-	−0.7562	—	1.00	—

orbitals singly occupied or a singlet if the lower of the two is occupied doubly. The energy gap of 0.8 ev is in a range where a clear prediction is not possible by the present crude model. However, for either the singlet or triplet ground state, the species should be colored as a result of a transition from a more deeply buried orbital to the lowest empty orbital (singlet case) or the two half filled orbitals (triplet case). In S_4 a similar situation obtains for a planar boat shape which is calculated to have an energy gap of 0.9 ev between the two orbitals available to the highest energy electrons. However, both stretched out chains (see Figure 11c) are calculated to have much smaller gaps between these two top orbitals. This fact suggests that the ground state for a stretched linear form is a triplet. A small energy gap between the orbitals available for the highest energy electrons persists as the chain lengthens from S_4 to S_8, and the calculated first absorption moves to longer wavelength, as shown in Figure 12. The linear compounds whose spectra have been identified (vide ante) confirm the red shift predicted by the calculations.

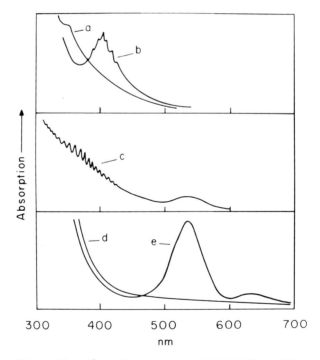

Figure 10.   Absorption spectrum of (a) S_3Cl_2 in iPC at 77°K; (b) S_3Cl_2 in iPC at 77°K after 5 minute photolysis; (c) S_3Cl_2 in Kr at 20°K after 1 minute photolysis; (d) S_4Cl_2 in iPC at 77°K, and (e) S_4Cl_2 in iPC at 77°K after 10 minutes photolysis

Polymeric sulfur now presents an anomaly. It is thought to consist of chains of length up to 500,000 S_8 units. It absorbs around 360 nm (4) and is yellow. Extrapolation of our calculations on linear S_n chains lead us to predict highly colored diradical species. The observed spectra, however, fit on the asymptotic limit of the sulfane chains. We are tempted to suggest that the chains of polymeric sulfur are terminated perhaps by hydrogen which is drawn from trace impurities. If polymeric sulfur contains an average of 5×10^5 atoms (33), only 10^{-5} weight of hydrogen or 10^{-4} weight of a hydrocarbon impurity like heptane is needed to supply enough hydrogen to saturate the chains. Since the purest sulfur known is about 99.999% pure, it can contain easily enough impurity to supply the hydrogen. Alternatively, the sulfur chains somehow may form closed rings or perhaps be a mixture of closed rings and sulfanes. We do not think an open chain structure is likely.

To complete our calculations we considered branched chain structures. As might have been expected, the sulfur at the branch point is calculated to have a high positive charge. Thus, branched S_4 is similar to SO_3. Calculations give too large an energy gap between top filled and lowest empty orbitals for a diradical species. However, branched S_4 is predicted to be colored, while branched S_5 is predicted to be colorless. The calculated charge distributions are shown (31). While the charge distribution of branched S_4 suggests a resonance structure,

that of S_5 defies simple representation. For the cyclic sulfur allotropes and for the sulfanes there is little charge build up. However, the linear sulfur allotropes show negative ends corresponding to structures such as:

We have looked at the sum of the extended Hückel orbital energies as a means of estimating relative stabilities of the various allotropic

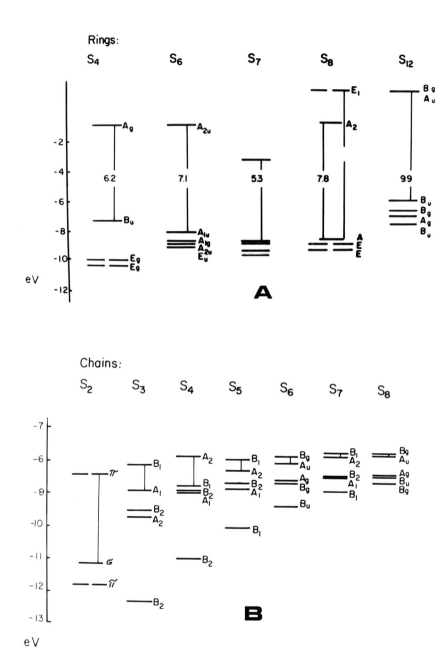

Figure 11.   *Energy levels of (a) sulfur rings, (b) sulfur chains, (c) isomers and conformers of S₄, and (d) sulfanes. All energy levels were computed with an*

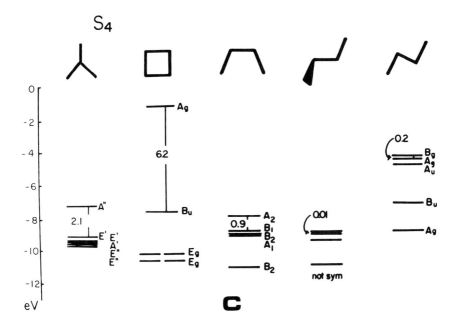

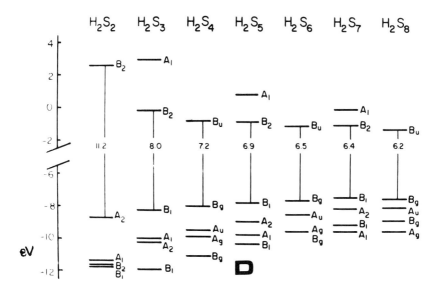

extended Hückel program based on the structural data indicated and with parameters listed in Ref. 31.

species. Figure 11c shows the results for several S_4 species. The total energies of the occupied orbitals are in the order branched < boat < helical < cyclic < chains. The colored nature of S_4 that we prepared from S_4Cl_2 and its agreement with the observed vapor spectra (vide ante) suggest that the normal vapor form is linear. However, the calculations suggest that branched S_4 is a possible allotrope. For S_5 the calculations show that the branched form has higher total energy than the linear and that this structure is probably unfavorable.

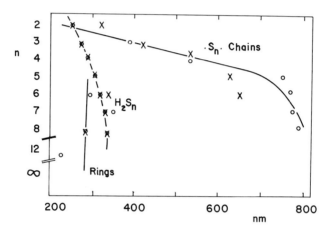

Figure 12.   Energy gap of the first allowed transition vs. the number of sulfur atoms in the molecule; (a) rings, (b) diradical chains, (c) sulfanes. O signifies calculated value, X signifies observed value.

Our calculations lead to the following general conclusions about sulfur allotropes:

(1) The rings should have large gaps between top filled and lowest empty orbitals and are pale yellow or colorless.

(2) The chains are colored highly with the color shifting to the red as the chain length increases; except for S_3 and boat S_4 the ground states are probably biradicals.

(3) The sulfanes are colorless with the absorption edge approaching an asymptote around 400 nm as the chain lengthens.

(4) Polymeric sulfur is probably a terminated chain, likely a sulfane; possibly it is cyclic, but it is unlikely to be an open chain.

(5) Vapor S_4 is not a ring; it is likely a boat or a helical chain; an S_4 allotrope similar in form to SO_3 is possible.

The extended Hückel calculations allow a fairly coherent qualitative interpretation of sulfur spectra. They suggest that more refined semi-empirical calculations and perhaps ab initio calculations on series such as the five S_4 allotropes yield interesting results.

Acknowledgment

This work was supported partially by the National Science Foundation Grant GP-9234, the National Air Pollution Control Agency Grant AP 639-02, and the Environmental Protection Agency Grant 639-03. We thank E. Davidson for the use of his extended Hückel calculation program. We also thank B. Gotthardt for help with preparing compounds and running spectra. Sections 1 and 2 are based on results obtained by B. Meyer and T. V. Oommen; section 3 is based on experiments of B. Meyer, T. V. Oommen, and T. Stroyer-Hansen; section 4 is based on work by B. Meyer, B. Gotthardt, D. Jensen, and T. V. Oommen; and section 5 contains work by K. Spitzer, M. Meyer, and M. Gouterman.

Literature Cited

(1) Meyer, B., *Chem. Rev.* (1964) **64**, 431.
(2) Clark, L. B., Ph.D. Thesis, University of Washington, Seattle (1963).
(3) Fukuda, M., *Mem. Coll. Sci. Kyoto Imp. Univ.* (1921) **4**, 351.
(4) Oommen, T. V., Ph.D. Thesis, University of Washington, Seattle (1970).
(5) Meyer, B., Oommen, T. V., *Sulphur Inst. J.* (1971) **6**, 2.
(6) Gilleo, M. A., *J. Chem Phys.* (1951) **19**, 1291.
(7) Bass, A. M., *J. Chem. Phys.* (1953) **21**, 80.
(8) Schmidt, M., Wilhelm, E., *Inorg. Nucl. Chem. Lett.* (1965) **1**, 39.
(9) Schmidt, M., Block, B., Block, H. D., Kopf, H., Wilhelm, E., *Angew. Chem. Int.* (1968) **7**, 632.
(10) Schmidt, M., Knippschild, G., Wilhelm, E., *Chem. Ber.* (1968) **101**, 381.
(11) Pratt, C. J., *Sci. Amer.* (1970) **222 (5)**, 63.
(12) Mondain-Monval, P., Job, R., Galet, P., *Bull. Soc. Chim.* (1930) **4 (47)**, 545.
(13) Harris, R. E., *J. Phys. Chem.* (1970) **74**, 3102.
(14) Meyer, B., Oommen, T. V., Jensen, D., *J. Phys. Chem.* (1971) **75**, 912.
(15) Srb, I., Vasko, A., *J. Chem. Phys.* (1962) **37**, 1892.
(16) Barrow, R. F., du Parcq, R. P., "Elemental Sulfur," B. Meyer, Ed., Interscience, New York, 1965.
(17) Graham, A., *Proc. Roy. Soc.* (1910) **84A**, 311.
(18) d'Or, L., *Compt. Rend. Acad. Sci. France* (1935) **201**, 1026.
(19) Preuner, G., Brockmöller, I., *Z. Phys. Chem.* (1913) **81**, 129.
(20) Berkowitz, J., "Elemental Sulfur," B. Meyer, Ed., Interscience, New York, 1965.
(21) Detry, D., Drowart, J., Goldfinger, P., Keller, H., Rickert, H., *Z. Phys. Chem.* (1967) **55**, 317.
(22) Wieland, K., Humpert, L., private communication, 1970.
(23) Braune, H., Steinbacher, E., *Z. Naturforsch.* (1952) **7a**, 486.
(24) Jensen, D., Ph.D. Thesis, University of Washington, Seattle (1971).
(25) Morelle, A., Ph.D. Thesis, University of Washington, Seattle (1971).
(26) Miller, D. J., Cusachs, L. C., *Chem. Phys. Lett.* (1969) **3**, 501.
(27) Feher, F., Munzner, H., *Chem. Ber.* (1963) **96**, 1131.
(28) Buttet, J., *Chem. Phys. Lett.* (1967) **1**, 297.
(29) Palma, A., Cohan, N. V., *Rev. Mex. Fis.* (1970) **19**, 15.
(30) Zerner, M., Gouterman, M., *Theor. Chim. Acta.* (1966) **4**, 44.
(31) Spitzer, K., Gouterman, M., Davidson, E., Meyer, B., to be published.
(32) Raymonda, J. W., Simpson, W. T., *J. Chem. Phys.* (1967) **47**, 430.

(33) MacKnight, W. J., Tobolsky, A. V., "Elemental Sulfur," B. Meyer, Ed., Interscience, 1965.
(34) Meyer, B., Oommen, T. V., Jensen, D., "Sulfur in Organic and Inorganic Chemistry," Vol. II, Chapter 12, A. Senning, Ed., Marcel Decker, New York, 1971.
(35) Moore, C. E., "Atomic Energy Levels," Vol. I, Circular of the U. S. National Bureau of Standards, 3467, 1958.

RECEIVED March 5, 1971.

5

Recent Advances in the Chemistry of Transition Metal Complexes of Unsaturated Bidentate Sulfur Donor Ligands (Metal Dithienes)

G. N. SCHRAUZER

The University of California, San Diego, Revelle College, La Jolla, Calif. 92037

The transition metal complexes of ligands derived from 1,2-dithiodiketones and/or cis-1,2-ethylenedithiols, known as metal dithienes or dithiolenes, undergo reactions typical of unsaturated organic compounds. This behavior is rationalized by topological considerations. The planar bisdithienes of Ni(II), Pd(II), and Pt(II) of composition $M[S_2C_2R_2]_2$, e.g., are closely related topologically to p-benzoquinone. The characteristic π-delocalization in metal dithienes accounts for the high electron affinity and electric conductivity of the neutral complexes. Dithiene resonance is also the essential stabilizing factor of trigonal prismatic geometry in the neutral trisdithienes, $M[S_2C_2R_2]_3$. Alkylation of anionic dithienes or dithiolatocomplexes affords new complexes—e.g., of composition $Ni[(CH_3)_2(S_2C_2Ph_2)_2]$. The latter reacts with $Fe(CO)_5$ to form a complex of diphenyl(thioketocarbene), $Ph_2C_2S[Fe_2(CO)_6]$. Synthesis of a number of macrocyclic complexes is described also.

Metal dithienes (*1*), which are known also as dithiolenes (*2*), comprise a group of neutral complexes of metals with ligands derived from 1,2-dithiodiketones (*1*) or *cis*-1,2-ethylenedithiols (*2*). The complexes

S   S
‖   ‖
R—C—C—R

1

H H
| |
S S
| |
R—C=C—R

2

interest coordination chemists because of their unusual properties and structures. The first representative of this class of complexes, the nickel-bisdithiene (3), was obtained initially as the product of the reaction of nickel sulfides with diphenylacetylene (Equation 1) (3):

$$NiS_x + Ph\text{—}C\equiv C\text{—}Ph \rightarrow$$

$$(1)$$

3

Complexes of most transition metals have become accessible by improved methods of synthesis. The most general method utilizes acyloins as starting materials which react first with P_4S_{10} to afford thiophosphates of substituted cis-1,2-ethylenedithiols. The latter react with metal salts in aqueous solution to give metal dithienes (4).

The dithienes are remarkably stable to acids and bases as well as thermally (on heating decomposition occurs in many cases only above 300°C). The dithienes resemble organic unsaturated compounds and, therefore, are described as highly delocalized. Initial confusion arose, however, because some authors believed in the ability of cis-1,2-ethylene-dithiols to stabilize high positive valence states of transition metals. The neutral bisdithiene (3), e.g., which is a Ni(II) derivative, was regarded for some time to be a Ni(IV) derivative because the ligands were assumed to be cis-1,2-ethylenedithiolatodianions rather than in an intermediate state between dithiolatodianions and neutral dithiodiketones. Although the concept of the π-delocalized nature of the complexes now is generally accepted, there is still considerable misunderstanding of how the ground state electronic configuration is formulated. Some authors have abandoned the metal oxidation state formalism and assume the valence state of the metal in the dithienes to be undefinable or fluctuating. This misconception has led to the rejection of the oxidation number formalism in other complexes. We will discuss first the electronic structure and bonding of metal dithienes from a topological point of view which encompasses the description of the bonding in the molecular orbital and valence bond approximations. We will outline the organic reactions and semiconductor properties of metal dithienes. This paper demonstrates the similarity of the complexes with organic delocalized systems—i.e., quinones—which makes them unique coordination compounds; however, the subject matter is not treated extensively. Thus, little if anything is said about dithienes of iron and cobalt, for example, and the complexes of the nickel and chromium group elements are emphasized. For a detailed account of earlier work References 1, 2, 5 should be consulted.

Topology and Electronic Structure of the Free Ligands

Unsaturated ligands are distinguished as even and odd systems depending on the number of vertical π-orbitals present (5). Odd sulfur ligands such as dithiocarboxylates or derivatives of 1,3-dithiodiketones form monoanions which as ligands show little residual electron affinity. Complexes of these odd ligands accordingly will behave normally and are described by the classical formalism (5). The ligands in a planar bis-complex of such ligands are regarded as independent of each other to a first approximation. In even ligands, such as the 1,2-dithiodiketones, low lying π-MO's are available and cause high electron affinity of these systems. In metal complexes it is therefore *a priori* difficult to decide if the ligands are present as the neutral 1,2-dithiodiketones, the *cis*-1,2-ethylenedithiolato dianions, or a state between the two extremes. Also, molecular orbitals extending over the whole molecule of the complex may be constructed because of the symmetry of the relevant MO's of the isolated ligand system and the metal p and d orbitals. This ambiguity of the even ligand systems is responsible for the difficulties in formulating the classical structures of the metal dithienes. Experimental complications arise from the instability of the isolated ligands, particularly of the 1,2-dithiodiketones. The reaction product of hexafluorobutyne with sulfur— *e.g.*, $(CF_3)_2C_2S_2$—does not exist noticeably as the 1,2-dithiodiketone but rather as the heterocycle 4, which is in equilibrium with a dimeric species

<u>4</u>

(6). Aryl substituted dithiodiketones generated in solution condense rapidly to 1,4-dithiins and other products (1, 7). It is not readily possible to relate the properties of complexes suspected to contain 1,2-dithiodiketone ligands to observable properties of the free ligands. However, under special conditions protonated derivatives of radical cations of *cis*-1,2-ethylenedithiols are generated and have been characterized by esr measurements (*see* below, and Reference 8, loc. cit.).

Electronic Structure and Bonding in d^8 *Metal Bisdithienes*

The ground state electronic configuration of d^8 metal bisdithienes of composition $M[S_2C_2R_2]_2$ (symmetry D_{2h}) has been discussed elsewhere (1, 5, 9) but is treated briefly now to permit an extension of our description to related π-electron systems. The two butadiene-like neutral dithio-diketone ligands form a set of π-MO's transforming as $2B_{1u}$, $2B_{2g}$, $2A_u$, and $2B_{3g}$. The lowest unoccupied ligand MO's transform as B_{1u} and B_{2g}. Interaction of the orbitals of symmetry B_{1u} with the metal p_z orbitals yields a bonding MO which is occupied by the two valence electrons of the central metal. The B_{2g} ligand MO's interact with the metal d_{xz} orbital and $2b_{2g}$ becomes weakly antibonding. This orbital remains empty in the neutral dithiene but is occupied on reduction to the mono- and dian-ions. In the neutral d^8 metal dithienes the ligand adopts a state interme-diate between dithiolatodianions and 1,2-dithiodiketones, a situation adequately represented by the two limiting valence bond structures **5** and **6**, which we are using because of the topological significance of

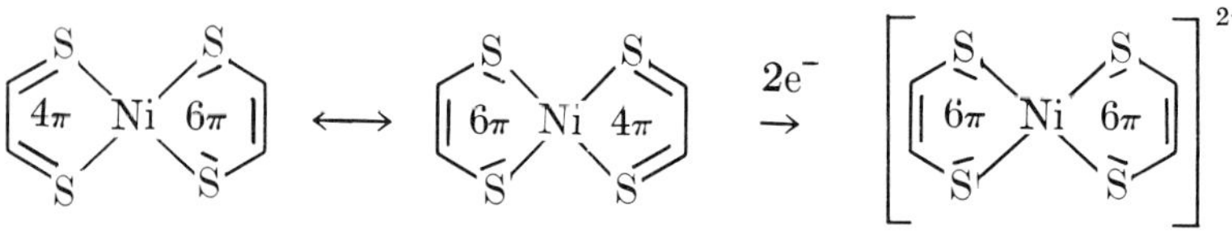

$$\text{(2)}$$

5   **6**   **7**

classical valence theory. The central metal, accordingly, is in the 2+ state of oxidation. On reduction to the dianion the electrons are placed into the significantly ligand-based $3b_{2g}$ MO. The ligands are converted into ethylenedithiolatodianions (2), which are formulated classically (Equation 2). The oxidation state of the metal is not changed during similar reduction reactions. The reaction resembles the reduction of p-benzoquinone to the hydroquinone dianion *via* the semiquinone radical anion as intermediate:

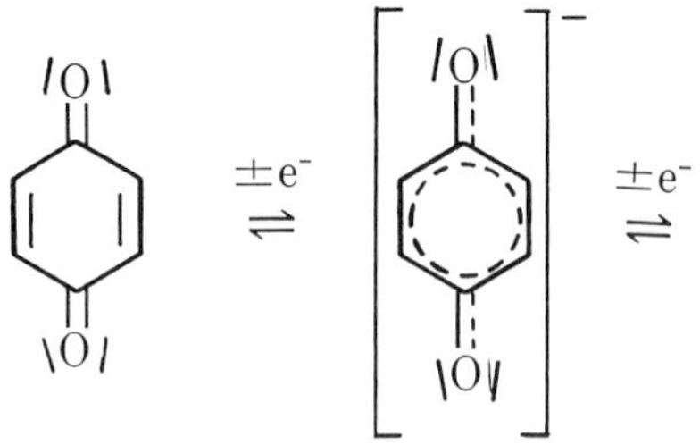

$$\text{(3)}$$

8   **9**   **10**

This apparent analogy is justified by a topological comparison of the π-electron system of the metal dithiene with that of the quinone. The vertical carbon and oxygen $2p$ orbitals of p-benzoquinone (symmetry

D_{2h}) transform as $3B_{1u}$, $3B_{2g}$, A_u, and B_{3g}. The 8 ligand C and S π-orbitals of the dithiene transform as $3B_{1u}$, $3B_{2g}$, $3B_{3g}$, and $2A_u$. Hence, all symmetry orbitals of the p-benzoquinone π-electron system are contained in the corresponding system of the d^8 metal bisdithiene. Also, the energy sequence of the π-MO's is similar in both systems (Table I). In both systems the lowest unoccupied π-MO has symmetry B_{2g} and is virtually nonbonding and responsible for the high electron affinity. Therefore, it is understood readily why metal dithienes exhibit substituent dependencies of the polarographic reduction potentials which are almost identical with those of p-benzoquinone derivatives. The analogy is substantiated further by the recent successful reduction of $Ni[S_2C_2Ph_2]_2$ to the hydride probable structure 11, the analog of p-hydroquinone (10):

$$Ni[S_2C_2Ph_2]_2{}^\circ \underset{TCNQ}{\overset{Zn\ dust/HOAC}{\rightleftarrows}} \quad \text{(11)} \tag{4}$$

3   11

The compound is relatively unstable and is oxidized back to the parent neutral dithiene by oxidizing agents such as tetracyano-p-benzoquinone.

Electronic Structure and Bonding in Trigonal Prismatic Trisdithienes

The trigonal prismatic geometry of the neutral trisdithienes $M[S_2C_2R_2]_3$ (11, 12) with M = Re, Mo, V, Cr, and W introduces a formal but unessential difficulty of topological classification with the organic π-electron systems. Chemically, the neutral trisdithienes resemble closely the planar bisdithienes of d^8 metals. The similarity in the chemical properties is obvious from the comparison with the published (1) schematic molecular orbital energy level diagrams of planar bis and trigonal prismatic trisdithienes. The π-MO's of the isolated dithiodiketone ligands are no longer degenerate in a trigonal prismatic arrangement because of developing interligand π-orbital interactions. This causes stabilization of the A'_2 and A''_1 MO's, but while the dithiodiketone ligands remain neutral, this type of interaction does not produce any stabilization of the trigonal prismatic geometry. Similarly, if all three ligands were present as cis-ethylenedithiolatodianions, no stabilization of the trigonal prismatic geometry would occur. However, the MO model calculations make it

Table I.   Symmetry Correlation and Energy Sequence of Molecular Orbitals of *p*-Benzoquinone and a Planar Bisdithiene Complex (Energy Not to Scale)

p-Benzoquinone   *Planar Bisdithiene*

p-Benzoquinone	Planar Bisdithiene	
$3b_{2g}$		
$3b_{1u}$	$3b_{1u}$	
	$3b_{3g}$	Antibonding
$1a_u$	$2a_u$	
$2b_{2g}$	$3b_{2g}$	
	$2b_{3g}(d_{yz})$	
	$2b_{2g}(d_{xz})$	Nonbonding
$2b_{1u}$	$2b_{1u}$	
	$1a_u$	
b_{3g}	$1b_{3g}$	Bonding
$1b_{2g}$	$1b_{2g}$	
$1b_{1u}$	$1b_{1u}$	

reasonable to assume that metal electrons will occupy the E' MO (*13*). In the neutral complexes of Cr, Mo, or W, for example, four electrons derived from the metal will occupy E', leading to a closed shell electronic configuration. The interaction of E' with metal orbitals of the same symmetry stabilizes E' and the trigonal prismatic geometry. In the neutral vanadium trisdithiene three metal electrons originally occupy E'. This condition should give rise to a Jahn–Teller-type distortion of the trigonal prismatic geometry. In the neutral Re trisdithiene one electron is placed into an antibonding MO, either $2a'_2$ or $5e'$. The trigonal prismatic structure still would be stabilized here, but the more electrons are placed into antibonding MO's, the more likely that isomerization to octahedral or distorted octahedral structure will occur. The fully reduced anionic trisdithienes are expected to be octahedral. The neutral trisdithienes resemble the planar d^8 metal dithienes since the orbital $4e'$ is similar in composition to the $2b_{1u}$ MO and since the unoccupied MO's $5e'$ and $2a'_2$ have energies similar to the $3b_{2g}$ orbital in the planar bisdithienes. The neutral trisdithienes of Cr, Mo, and W therefore are represented classically as resonance hybrids of the three main contributing limiting (Kekulé-type) structures (**12, 13, 14**) (*1, 5, 13*):

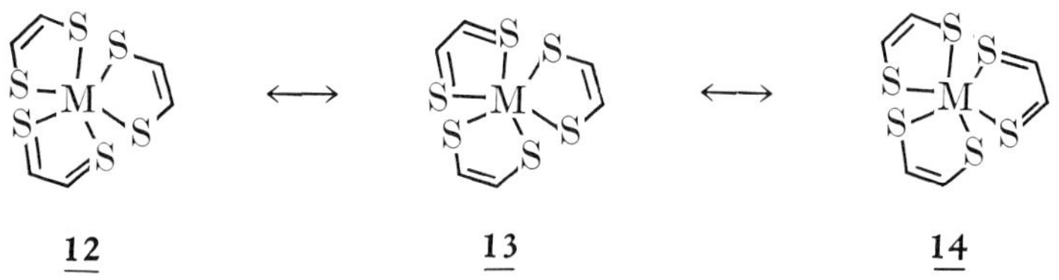

$$\tag{5}$$

12   **13**   **14**

Upon reduction to the anionic species the first two electrons occupy orbital $5e'$ or $2a'_2$. It is difficult to predict which orbital will be occupied because of similar orbital energies. This question is not important because of the change of coordination geometry following the addition of the two electrons. The ligands are converted to dithiolatodianions in this process, as in the reduction of the d^8 metal dithienes. The trisdithiene dianions accordingly are represented classically as trisdithiolato com-

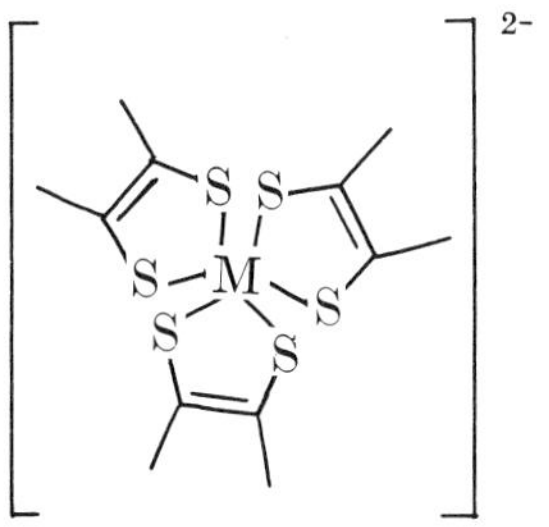

15

plexes (**15**), again demonstrating a close analogy to the quinones' behavior.

Organic Reactions of Metal Bis and Trisdithienes

Alkylation. The topological analogies outlined in the previous section are substantiated further by considering the chemistry of metal dithienes. We first discuss the alkylation of anionic species. This reaction is equivalent to the alkylation of the hydroquinone dianion and affords members of a class of new coordination compounds (*14*). With simple alkyl halides the d^8 metal dithiene dianions afford the 1,4-S-dialkylderivatives (**16**). With α,ω-dibromoalkanes at high dilution new chelates of type **17** were obtained for $x = 5$–12 (*3*). With smaller values of x or

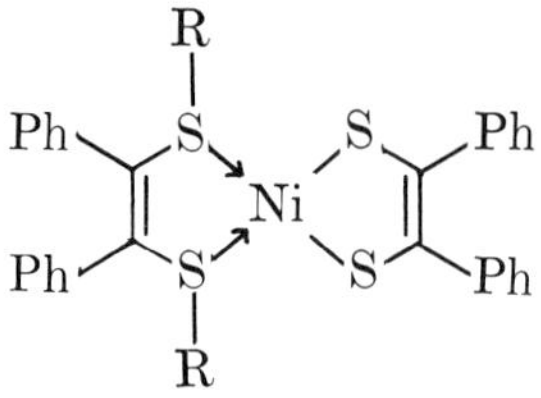

16 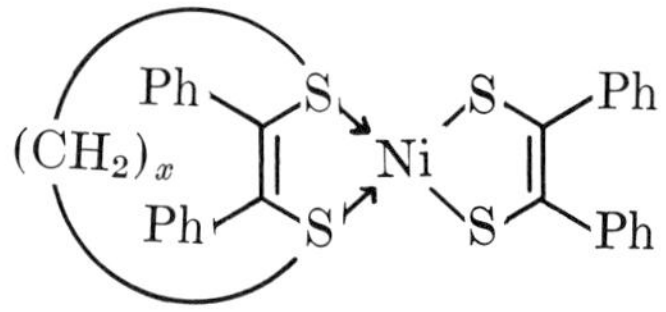   17

with excess alkylating agent, the corresponding 1,4-S-dihaloalkyldithio-lates are formed. These experiments recall Lüttringhaus' (*15*) experiments on the alkylation of hydroquinone with α,ω-polymethylene dibromides. The minimum value of x is 8 for the formation of the cyclic ether (**18**):

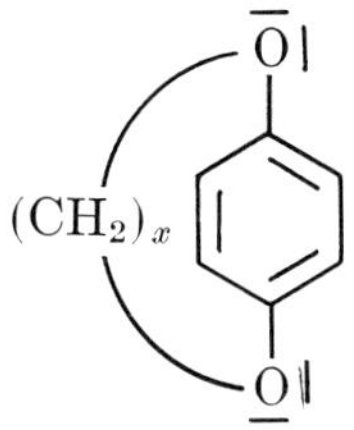

18

The neutral complex $Ni[(CH_3)_2S_2C_2Ph_2][S_2C_2Ph_2]$ reacts with excess methyl iodide on heating in a sealed tube to yield a mixture of nickel iodide and a presumably octahedral iodide of Ni(II) with three molecules of bis(*cis*-methylthiostilbene) as the ligands. The salt-like complex hydrolyzes rapidly (*10*).

The methylation of dianions of trisdithienes of group VI transition metals (*16*) provides an isolable 1,4-S-dimethyl derivative of composition $W[(CH_3)_2(S_2C_2Ph_2)_3]$ (**19**). The corresponding derivatives of Cr, Mo,

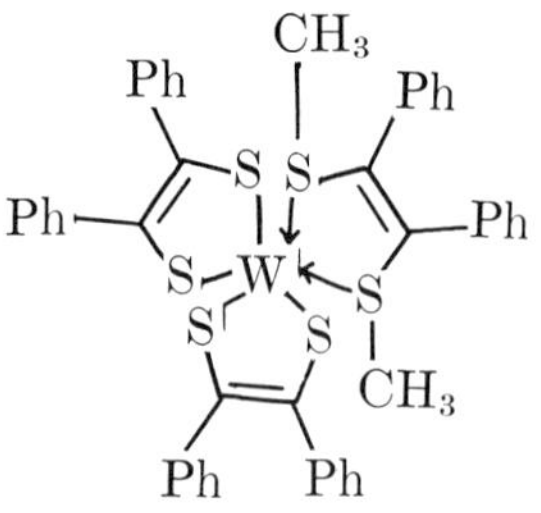

19 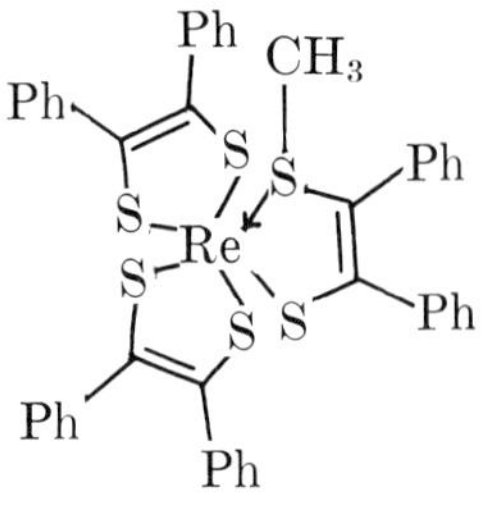   20

or V could not be isolated. However, the alkylation of the dianions of
these dithienes yields bis(*cis*-methylthiostilbene), indicating that methylation of the complex anions took place but that the complexes
$M[(CH_3)_2(S_2C_2Ph_2)_3]$ (M = V,Cr,Mo) are too unstable to be isolated
under the reaction conditions employed (16). The alkylation of anions
of iron and cobalt dithienes similarly afforded only alkyl derivatives of
the free *cis*-ethylenedithiols.

The infrared spectrum of 19 is typical of a 1,2-ethylenedithiolato
complex rather than of a dithiene and, thus, permits the classical formulation. This is consistent with the absence of typical dithiene reactivity.
Complex 19 does not undergo reversible electron transfer reactions, for
example (16).

The methylation of $Re[S_2C_2Ph_2]_3^-$ gives a purple monomethyl derivative (20), $Re[(CH_3)(S_2C_2Ph_2)_3]$, which possesses all properties of
a dithiolate. The existence of 19 confirms the ground state description
of the neutral group VI metal trisdithienes (12, 13, 14) and shows that
the complexes are electron deficient and two electrons short of the trisdithiolato structure (15). With these conclusions and the results of the
methylation experiment, we formulate the neutral trisdithiene of rhenium
as a resonance-hybrid of the three limiting structures, 21, 22, 23:

$$\tag{6}$$

21   22   23

The structures of 19 and 20 are unknown as yet. They could be trigonal
prismatic, but we consider them to be octahedral or distorted octahedral
because of their classical dithiolato structures (16).

Diels–Alder Reaction and the Formation of Adducts with Olefins.
The d^8-metal bisdithienes react with alkynes to form 1,4-dithiins (24)
according to Equation 7 (17):

$$\tag{7}$$

3   24   25

Equation 7 is related to a Diels–Alder type reaction and led us to examine the reaction of metal dithienes with reactive olefins. With norbornadiene (bicyclo[2.2.1]hepta-2,5-diene) in homogeneous solution (CH_2Cl_2 is a convenient solvent), the Diels–Alder adduct **26** is formed in high yield with the oligomeric nickel(II)-stilbenedithiolate (**25**) (*16, 17*):

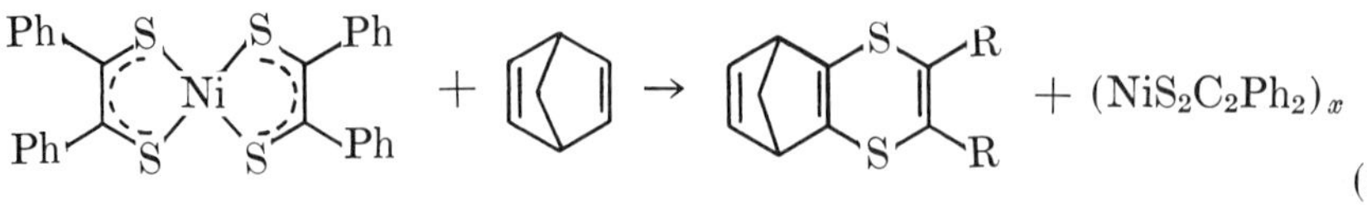

$$\tag{8}$$

   3   **26**   **25**

This reaction, which is allowed orbitally, is contrasted with the formation of a norbornadiene adduct by a symmetry forbidden reaction where the dithiene reacts presumably in its first excited state (**27**):

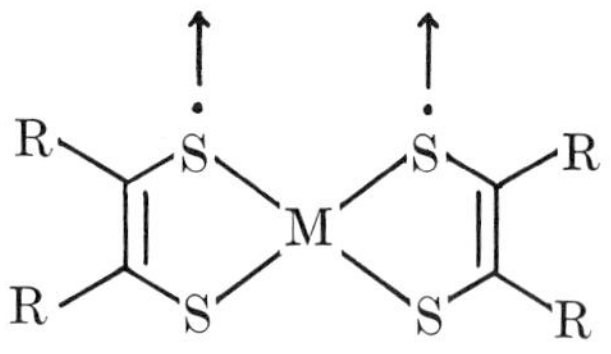

27

On refluxing **3** in norbornadiene in the absence of a solvent, the formation of **26** is suppressed, and the adduct $Ni[S_2C_2Ph_2]_2 \cdot C_7H_8$ (**28**) is formed preferentially. The adduct is stable thermally but on exposure to light decomposes into **3** and norbornadiene. A similar addition compound

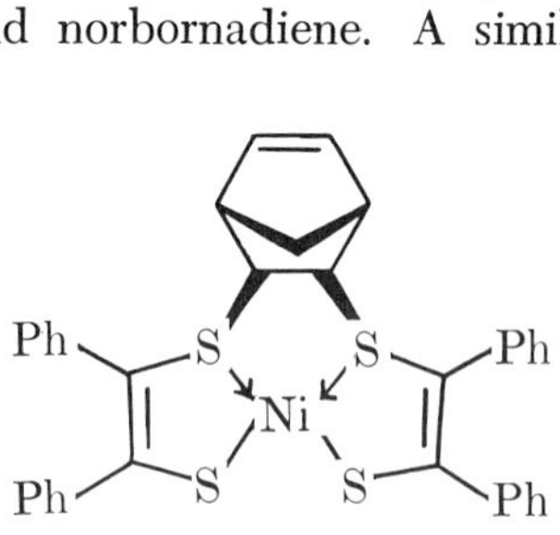

28

of norbornadiene was reported by Wing *et al.* (*18*), using $Ni[S_2C_2$-$(CF_3)_2]_2$ as the parent dithiene, whose structure was provided by x-ray analysis. These authors explored originally the reaction of quadricyclene with $Ni[S_2C_2(CF_3)_2]_2$, expecting the latter to catalyze its conversion to norbornadiene. Although some catalysis of this process was observed, the adduct formation could not be prevented (*18*). The structure of **28** was established (*16*) by chemical methods (reaction with CH_3I to **29**, followed by characterization of the thioether, **30**):

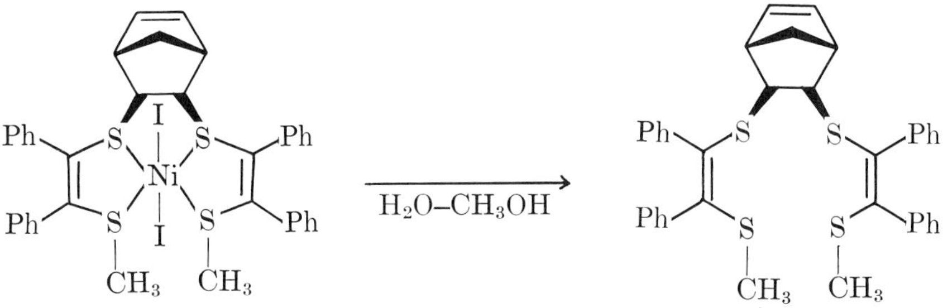

$$+ Ni^{2+} + 2I^- \qquad (9)$$

Reaction of **28** with α,α'-dibromo-*o*-xylene in toluene solvent afforded the macrocyclic complex (**31**), which hydrolyzes readily to yield **32** (*19*):

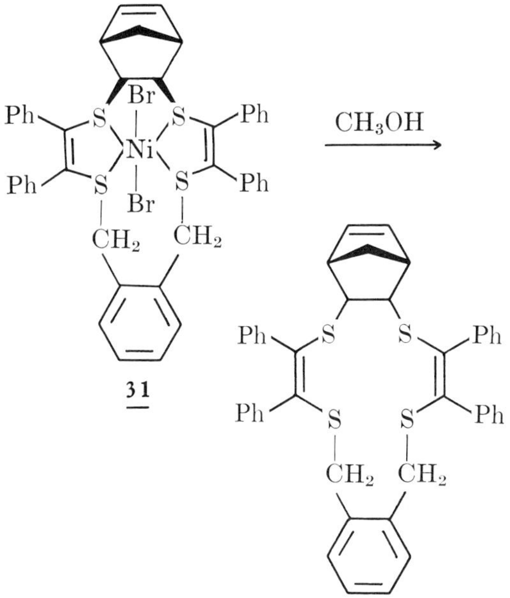

$$+ Ni^{2+} + 2Br^- \qquad (10)$$

The olefin adduct formation appears to be limited to the dithienes of Ni, Pd, and Pt. The formation of the Diels–Alder adduct **26** occurs on the reaction of $Mo(S_2C_2Ph_2)_3$, $Cr(S_2C_2Ph_2)_3$, and $V(S_2C_2Ph_2)_3$ but not of $W(S_2C_2Ph_2)_3$, $Re(S_2C_2Ph_2)_3$, or with various neutral or anionic iron and cobalt dithienes under comparable conditions. In the reaction of $Mo(S_2C_2Ph_2)_3$ with norbornadiene only one dithiene ligand is reactive, and a complex of composition $[Mo(S_2C_2Ph_2)_2]_x$ is formed. The molybdenum bisdithiene is presumably a cluster complex; molecular weight determinations suggest a hexameric structure. Upon heating, the complex decomposes to form $Mo(S_2C_2Ph_2)_3$ and molybdenum containing products (*20*).

Stabilization of a 1,3-Thioketocarbene through π-Complex Formation

The neutral $Ni[S_2C_2Ph_2]_2$ (*3*) reacts with iron pentacarbonyl to produce **33** (*21*). To prove the structure of the 1,4-S-dialkyl(stilbenedithiolato)-Ni(II) complex by the reaction with $Fe(CO)_5$, a complex (**36**) of composition $Ph_2C_2SFe_2(CO)_6$ was obtained (Equation 11) (*22*):

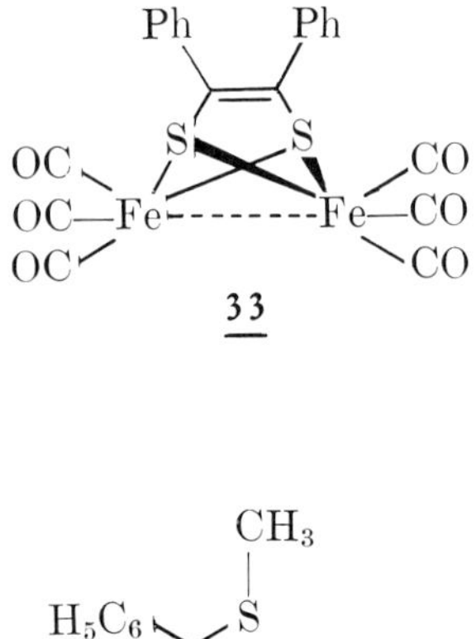

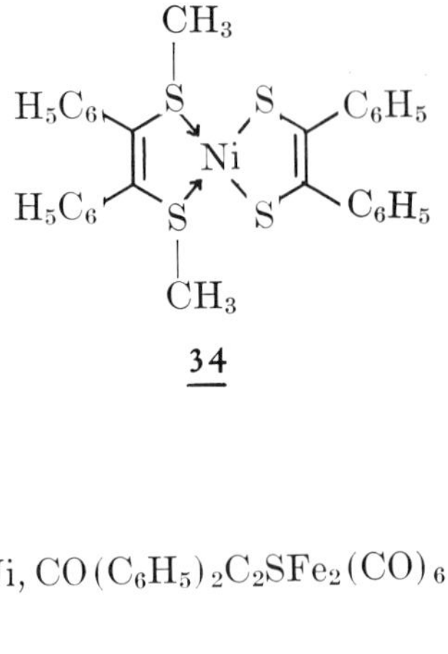

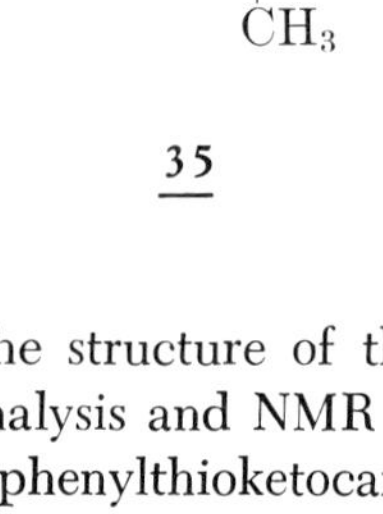

$$+ FeS, Ni, CO(C_6H_5)_2C_2SFe_2(CO)_6 \qquad (11)$$

The structure of the complex was elucidated by x-ray crystallographic analysis and NMR spectroscopy, and it was shown to be a π-complex of diphenylthioketocarbene, Ph—C—C(=S)—Ph (Figure 1) (*22*):

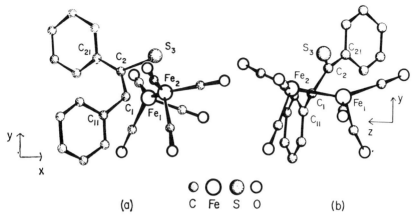

Journal of the American Chemical Society

Figure 1. The structure of the $(C_6H_5)_2C_2SFe_2(CO)_6$ complex viewed along (a) the z axis and (b) the x axis (22).

Thioketocarbenes are 1,3-dipoles whose ground state are represented (23, 24) by the limiting structures **37** and **38**. In complex **36** the thio-

$$\text{H}_5\text{C}_6\!-\!\overset{\displaystyle \overset{\text{S}}{\|}}{\text{C}}\!-\!\ddot{\text{C}}\!-\!\text{C}_6\text{H}_5 \longleftrightarrow \text{H}_5\text{C}_6\!-\!\overset{\displaystyle \overset{\text{S}^-}{|}}{\text{C}}\!=\!\overset{+}{\text{C}}\!-\!\text{C}_6\text{H}_5 \qquad (12)$$

$$\textbf{37} \hspace{5.5cm} \textbf{38}$$

ketocarbene ligand behaves as the electronic structure suggested by the limiting VB formulation. One of the iron atoms (Fe_1) is bonded symmetrically to the C—C—S system. The bond between Fe_2 and C_1, however, is shorter than that between Fe_2 and C_2, suggesting a considerable degree of σ-bond character. The bifunctionality of the ligand in **36** is consistent with the intermediate thioketocarbene-1,3-dipolar structure. The bonding situation is shown in Figure 2 (22).

Electric Properties of Metal Dithienes

The neutral dithienes $M[S_2C_2R_2]_{2\text{ or }3}$ and related compounds are semiconductors comparable with unsaturated hydrocarbons and charge-transfer complexes. Observed (25) resistivities at room temperature are 10^3–10^{15} ohm-cm, depending on the structure of the complexes, the central metal, and the nature of the substituents, R. A linear correlation between the resistivity and the first polarographic half-wave potential of reduction was observed for the planar d^8 metal bisdithienes $M[S_2C_2R_2]_2$ (M = Ni,Pd,Pt) for series with the same substituent, R. There is also a correlation between the energy of the first intense π–π^* transition in the near infrared region and the resistivity, suggesting that the electrons are transported in the lowest unoccupied π-MO during electric conduction. Hall-effect measurements identify the majority of charge carriers as negative. The conductivity decreases with increasing size of the substituents, R, as expected, although electronic and inductive effects of the substituents sometime counteract size effects. For nickel complexes the order of decreasing conductivity is H, Ph, CH_3, p-$C_6H_4CH_3$, and p-$C_6H_4OCH_3$ (25). A similar sequence was observed for the palladium complexes, but for the platinum derivatives it is CH_3, Ph, p-$C_6H_4OCH_3$, and p-$C_6H_4CH_3$. The dithiene with the highest reported bulk conductivity of 10^{-3} ohm^{-1}-cm^{-1} is $Re(S_2C_2H_2)_3$. Single crystal measurements reveal a nearly isotropic conductivity for $Ni[S_2C_2Ph_2]_2$ and $Re[S_2C_2(CH_3)_2]_3$. The prevalent mechanism of electric conduction is the injection of electrons from the cathode into the lowest unoccupied π^* MO of the complexes ($3b_{2g}$ for planar d^8 metal bisdithienes). In $Ni[S_2C_2H_2]_2$ the nearest sulfur atoms are separated by 3.04 A. In $Ni[S_2C_2Ph_2]_2$ the distance between the nearest sulfur atoms is 4.6 A. The first electronic excitation in the former is observed at 720 mμ, in the latter at 866 mμ, but the resistivities of the two complexes are 1×10^7 and 7×10^7 ohm-cm, respectively, indicating the effect of changing intermolecular distances.

Considerable carrier trapping was observed in measurements of the thermo-electric power coefficient (25). The complex $Ni[S_2C_2Ph_2]_2$ was exposed consequently (in polycrystalline form) to a 1.5 Mev electron

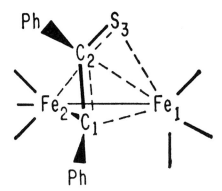

Figure 2. Bonding of the diphenyl-thioketocarbene to the 2 $Fe(CO)_3$ groups in Complex 36. Observed (22) bond distances are Fe_1-C_1, 2.089 (6); Fe_1-C_2, 2.061 (6); Fe_1-S_3, 2.243 (2) A, respectively. Fe_2-C_1 is 1.969 (7); Fe_2-C_2, 2.654 (6); Fe_2-S_3, 2.256 (2) A (standard deviations in parentheses). The angle C_1-C_2-S_3 is 103.3 (5) A.

beam. This produced a potential which persisted for several weeks after exposure and which could not be discharged by shorting the circuit. This potential suggested the presence of immobile charged species—*i.e.*, Ni[S$_2$C$_2$Ph$_2$]$_2$—which cannot be discharged at reasonable rates (25).

The resistivities of metal dithienes compare with those of Ovalene ($\rho = 2.3 \times 10^{15}$ ohm-cm), violanthrone (2.3×10^{10} ohm-cm), or cyananthrone (1.2×10^7 ohm-cm) (19). Metal phthalocyanines with reported resistivities of between 10^{11} and 10^{15} ohm-cm, are worse semi-conductors than the dithienes. The conductivity of some rhenium dithienes approaches that of intrinsic silicon; even better conductors may result by synthesizing polymeric species.

Donor–Acceptor Complexes

The CF$_3$ substituted nickel dithiene (occasionally referred to as nickel "dithiete"), Ni[S$_2$C$_2$(CF$_3$)$_2$]$_2$, has probably the highest electron affinity of all known neutral nickel dithienes and was found to form 1:1 donor–acceptor complexes with perylene and pyrene (26). The crystal structures of these adducts were determined from single crystal x-ray diffraction data and revealed them to consist of alternating stacks of the hydrocarbon and the dithiene. The interplanar spacing in both complexes is 3.54 A. All available evidence suggests that the ground-state of the complexes is neutral rather than ionic. Although a weak ESR signal is observed usually, it is presumably a result of an impurity. The complexes are semi-conductors with room temperature bulk resistivities of 10^5 ohm-cm. The conductivity is considerably less than reported for the perylene–iodine charge-transfer complex ($\rho = 1.0 \times 10^1$ ohm-cm), which presumably has much greater ionic contributions to the bonding in the ground state. The complexes exhibit charge-transfer bands in the near infrared region at 4000 cm^{-1} (perylene complex) and at 8.500 cm^{-1} (pyrene complex) (26). Similar electron donor–acceptor complexes of other dithienes have not been characterized yet.

Cation Radicals of Cis-1,2-Ethylenedithiols

In attempts to generate cation radicals of metal dithienes of composition M(S$_2$C$_2$R$_2$)$_2^+$ $_{or \; 3}$, the oxidation of neutral dithienes was studied at various conditions (8). Oxidation of Ni[S$_2$C$_2$Ph$_2$]$_2$ by Br$_2$ afforded NiBr$_2$, sulfur, and tetraphenyl-1,4-dithiin even when the reaction was conducted under mild conditions, suggesting that cation radicals of metal dithienes are unstable. Milder methods of oxidation were utilized—*e.g.*, reaction of the complexes with H$_2$SO$_4$ or AlCl$_3$ in nitromethane. Under these

Table II.   ESR Data of Cation Radicals

Substituent R		$\langle g \rangle$ (isotropic)[a]
CH_3	$Ni(S_2C_2(CH_3)_2)_2$	2.014
CH_3	$Pd(S_2C_2(CH_3)_2)_2$	2.014
CH_3	$Mo(S_2C_2(CH_3)_2)_3$	2.014
H	$Ni(S_2C_2H_2)_2$	2.014
C_6H_5	$Ni(S_2C_2(C_6H_5)_2)_2$	2.014
C_6H_5	$Pd(S_2C_2(C_6H_5)_2)_2$	2.014
$p\text{-}CH_3\text{-}C_6H_4$	$Ni(S_2C_2(C_6H_4CH_3)_2)_2$	2.014
$p\text{-}CH_3O\text{-}C_6H_4$	$Ni(S_2C_2(C_6H_4OCH_3)_2)_2$	2.013

[a] Average error in $\langle g \rangle$ = 0.002.
[b] In $H_2SO_4\text{-}CH_3NO_2$ glass at 100°K.

conditions oxidation of the dithienes is accompanied by solvolysis, affording cation radicals of cis-1,2-ethylenedithiols (39), and cation radicals of 1,4-dithiins (40) as by-products (8).

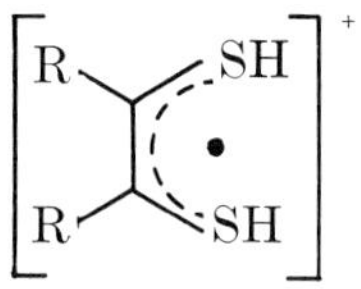

39 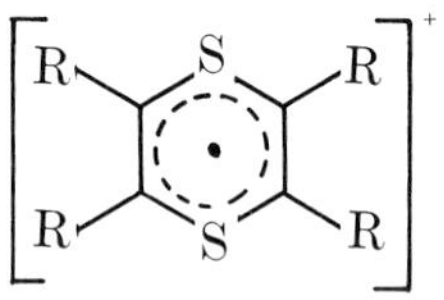   40

The radical cations were characterized by ESR measurements and chemical reactions. The cations with R=H and CH_3 exhibit hf splitting as a result of hydrogen, permitting the comparison of observed spin-densities with expectation values obtained from calculations by an iterative extended HMO method. For 39 with R=H the observed spin density on carbon of 0.122 agrees with the expectation value of 0.113. A characteristic of these radicals is the high spin density on sulfur (observed 0.378) and the three-fold anisotropy of the g-tensor (Table II) (8).

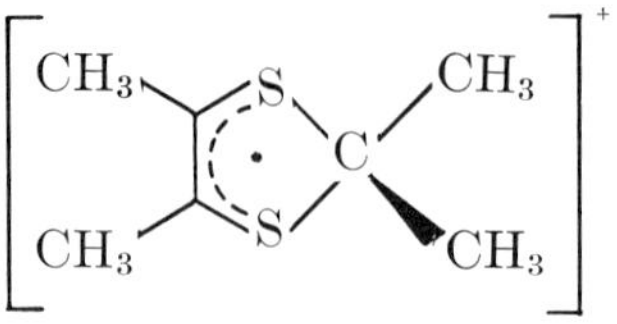

$g = 2.009$

$\langle \alpha_H \rangle = 5.75$ G

41

of *cis*-1,2-Ethylenedithiols (*8*)

$\langle a_H \rangle$, G	Anisotropy[b]			Hf splitting multiplicity
	$\langle g_1 \rangle$	$\langle g_2 \rangle$	$\langle g_3 \rangle$	
2.06	2.022	2.016	2.005	7
2.06	2.022	2.016	2.005	7
2.06	2.022	2.016	2.005	7
2.75	2.026	2.020	2.003	3
	2.025	2.020	2.003	
	2.025	2.019	2.003	
	2.020	2.018	2.003	
	2.023	2.019	2.003	

The cation radicals of type **39** condense to 1,4-dithiins and 1,4-dithiin radical cations, respectively. The *cis*-geometry of the radicals **39** was established by conversion into the radical cation **41**, whose identity was confirmed by independent synthesis. The apparent instability of cation radicals of dithienes in solution does not mean that such species are incapable of existence. The ion $Ni[S_2C_2(CH_3)_2]_2^+$ has been detected in the mass spectrum of the corresponding neutral dithiene (*11*).

Conclusion

The synthesis and properties of a number of bis(trifluoromethyl)di-selenenes (1,2-diselenolenes) $M[Se_2C_2(CF_3)_2]_2$, $M[Se_2C_2(CF_3)_2]_3$, and of the heterocycle $Se_2C_2(CF_3)_2$ have been reported by Davison and Shaw (*12*). The complexes are much less stable than the dithienes but otherwise exhibit similar properties and reactions. The blue–black $Ni[Se_2C_2(CF_3)_2]_2$ is reduced to the green mono- and the orange dianion. In the ESR spectrum of the monoanion ^{77}Se hyperfine splitting is resolved, suggesting 70–90% ligand character of the half-occupied MO more than in the corresponding orbital of the dithiene where the ligand character is approximately 50% (*12*). The crystal structure of $Mo[Se_2C_2(CF_3)_2]_3$ has been determined; the coordination geometry is trigonally prismatic as in the dithienes (unpublished work, cited in Reference 2).

Complexes of 1,2-dithiotropolone were predicted to possess properties different from metal dithienes, the ligand being a typical odd system (*1*). Forbes and Holm (*27*) reported recently the synthesis of the ligand and a number of complexes which confirm this proposal.

The electronic spectra of dithioacetylacetone complexes of Ni(II), Pd(II), and Pt(II) were analyzed on the basis of extended HMO calculations (*28*). Without considering the topological properties of this ligand, McCleverty (*2*) suggested that these complexes should have dithiene-like properties, in contrast with our own prediction (1). It is confirmed now that the dithioacetylacetonates of Ni, Pd, and Pt are interpreted in the conventional fashion owing to the impossibility of dithiene resonance in these odd ligand complexes. Tris-complexes of this ligand would be expected also to be octahedral rather than trigonal prismatic although structures of group VI transition metal complexes of this ligand are unknown. These final examples confirm the description of metal dithienes as a topologically distinct class of extensively delocalized coordination compounds, exhibiting close similarities to unsaturated organic systems—*e.g.*, the quinones. Their chemical properties and availability from inexpensive starting materials make them ideal for further study. The complexes attain also practical utility as efficient antioxidants and catalysts of hydroperoxide decomposition (29), as pigments for special applications (*i.e.*, protection against near infrared and ultraviolet radiation) or fuel additives.

Whereas the neutral bisdithienes $M[S_2C_2R_2]_2$ of Ni, Pd, and Pt are isomorphous with R = phenyl, indicating no intermolecular metal–metal interactions. A recent x-ray crystallographic analysis of the corresponding complexes with R=H shows that the Pd and Pt derivatives form discrete dimers (30). In the Pd complex dimer each metal atom is drawn inward from the plane of the four sulfur atoms by 0.12 A, leading to a Pd–Pd distance of 2.79 A. In the corresponding Pt complex dimer the Pt–Pt distance is 2.77 A; in both cases the metal–metal separations are comparable with those in the metal. In $Ni[S_2C_2H_2]_2$ no similar dimerization or metal–metal interaction is observed, indicating the greater relative importance of metal–metal bonding as compared with intermolecular S—S interactions. However, the formation of the dimers also involves bonding interactions between the ligands. Conceivably, the dimer formation is related topologically to the formation of quinhydrones.

Acknowledgments

I express my thanks and appreciation to my co-workers and colleagues who have participated in the work on metal dithienes. At Munich, V. P. Mayweg, W. Heinrich, and H. W. Finck. At Shell Development Co., E. J. Rosa, A. E. Smith, and D. C. Olson. At La Jolla, H. N. Rabinowitz, R. P. Murillo, and R. K. Y. Ho. Particular thanks are also extended to I. C. Paul (Urbana, Ill.) for confirming the structure of complex 36 by single crystal x-ray diffraction.

Literature Cited

(1) Schrauzer, G. N., *Transition Metal Chem.* (1968) 4, 299, and references cited therein.

(2) McCleverty, J. A., *Progr. Inorg. Chem.* (1968) **10**, 49; for additional reviews see H. B. Gray, *Advan. Chem. Ser.* (1967) **62**, 641 and L. F. Linday, *Coord. Chem. Rev.* (1969) **4**, 41.
(3) Schrauzer, G. N., Mayweg, V. P., *J. Amer. Chem. Soc.* (1962) **84**, 3221.
(4) Schrauzer, G. N., Mayweg, V. P., Heinrich, W., *Inorg. Chem.* (1965) **4**, 1615.
(5) Schrauzer, G. N., *Accounts Chem. Research* (1969) **2**, 72.
(6) Krespan, C. G., *J. Amer. Chem. Soc.* (1961) **83**, 3434.
(7) Schrauzer, G. N., Finck, H. W., *Angew. Chem.* (1964) **76**, 143.
(8) Schrauzer, G. N., Rabinowitz, H. N., *J. Amer. Chem. Soc.* (1970) **92**, 5769.
(9) Schrauzer, G. N., Mayweg, V. P., *J. Amer. Chem. Soc.* (1965) **87**, 3585.
(10) Murillo, R., Ho, R. K. Y., Schrauzer, G. N., unpublished work.
(11) Bloom, S. M., Dudek, G. O., *Inorg. Nucl. Chem. Lett.* (1966) **2**, 183.
(12) Davison, A., Shawl, E. T., *Inorg. Chem.* (1970) **9**, 1820.
(13) Schrauzer, G. N., Mayweg, V. P., *J. Amer. Chem. Soc.* (1966) **88**, 3235.
(14) Schrauzer, G. N., Rabinowitz, H. N., *J. Amer. Chem. Soc.* (1968) **90**, 4297.
(15) Lüttringhaus, A., *Naturwissenchaften* (1942) **30**, 40.
(16) Schrauzer, G. N., Rabinowitz, H. N., *J. Amer. Chem. Soc.* (1969) **91**, 6522.
(17) Schrauzer, G. N., Mayweg, V. P., *J. Amer. Chem. Soc.* (1965) **87**, 1483.
(18) Wing, R. M., Tustin, G. C., Okamura, W. H., *J. Amer. Chem. Soc.* (1970) **92**, 1935.
(19) Schrauzer, G. N., Ho, R. K. Y., Murillo, R., *J. Amer. Chem. Soc.* (1970) **92**, 3508.
(20) Ho, R. K. Y., Murillo, R. P., unpublished work.
(21) Schrauzer, G. N., Mayweg, V. P., Finck, H. W., Heinrich, W., *J. Amer. Chem. Soc.* (1966) **88**, 4604.
(22) Schrauzer, G. N., Rabinowitz, H. N., Frank, J. A. K., Paul, I. C., *J. Amer. Chem. Soc.* (1970) **92**, 212.
(23) Huisgen, R., *Angew. Chem.* (1963) **75**, 604, 741.
(24) Huisgen, R., *Angew. Chem. Int. Ed. Engl.* (1963) **2**, 565, 633.
(25) Rosa, E. J., Schrauzer, G. N., *J. Phys. Chem.* (1969) **73**, 3132.
(26) Schmitt, R. D., Wing, R. M., Maki, A. H., *J. Amer. Chem. Soc.* (1969) **91**, 4394.
(27) Forbes, C. E., Holm, R. H., *J. Amer. Chem. Soc.* (1970) **92**, 2297.
(28) Siimann, O., Fresco, J., *J. Amer. Chem. Soc.* (1970) **92**, 2652.
(29) Uri, N., *Israel J. Chem.* (1970) **8**, 125.
(30) Browall, K. W., Interrante, L. V., Kasper, J. S., *J. Amer. Chem. Soc.* (1971) **93**, 6289.

RECEIVED March 5, 1971.

6

The Structures of Sulfur–Nitrogen Compounds

WILLIAM L. JOLLY

Department of Chemistry, University of California, and Inorganic Materials Research Division, Lawrence Berkeley Laboratory, Berkeley, Calif. 94720

The structures and bonding of the following species have been studied: S_2N_2, $S_2N_2^+$, $S_2N_2^{2+}$, $S_3N_3Cl_3$, $S_3N_3O_3Cl_3$, S_7NH, $S_6(NH)_2$, $S_5(NH)_3$, $S_4N_4H_4$, $S_4N_4F_4$, $S_4N_4^{2-}$, S_4N_4, $S_4N_4^{2+}$, $S_5N_5^+$, $S_3N_2O_2$, $S_3N_2Cl^+$, $S_3N_3NPPh_3$, S_4N_2, and $S_4N_3^+$. These species are characterized by rings or chains of alternating sulfur and nitrogen atoms. The structures are rationalized by comparison with isoelectronic compounds of known structure and by assuming delocalized $p\pi$ bonding whenever possible.

In this paper the structures of compounds which contain rings or chains of alternating sulfur and nitrogen atoms are discussed. During the last 15 years or so this class of compounds has frustrated and mystified chemists because various structure determinations have often showed that structures which had been predicted for these compounds were wrong. However, enough structures of such compounds are now known that it is possible to systematize and rationalize the data. A discussion of this rationale and its use in making several predictions are presented below.

The structural rationale is based partly on the comparison of sulfur–nitrogen compounds with isoelectronic compounds of known structure and on the assumption of delocalized $p\pi$ bonding whenever possible. In the study of these structures, it was apparent that certain empirical bonding rules must be remembered; these are discussed in the following section.

Empirical Bonding Rules

Rule 1. BOND ANGLES AT SULFUR ATOMS ARE USUALLY IN THE NEIGHBORHOOD OF 100°. This rule is not peculiar to sulfur–nitrogen compounds.

For other examples we cite the Cl–S–Cl angle of 100.3° in SCl_2 and the F–S–F and O–S–F angles of 92.8 and 106.8°, respectively, in SOF_2 (*1*). The rule indicates the tendency for sulfur atoms to use mainly orthogonal p orbitals in bonding.

Rule 2. TERVALENT SULFUR ATOMS—*i.e.*, SULFUR ATOMS BONDED TO THREE OTHER ATOMS—CANNOT ENGAGE IN $p\pi$–$p\pi$ BONDING. This rule is reasonable in view of Rule 1. The lone pair of a tervalent sulfur atom is essentially in a 3*s* orbital, and extensive hybridization is required to use it in π bonding.

Rule 3. BOND ANGLES AT DIVALENT NITROGEN ATOMS WHICH ARE INVOLVED IN π BONDING ARE USUALLY 120° OR GREATER. The larger bond angles at nitrogen atoms are probably a result of the smaller size of nitrogen atoms and the greater repulsion between valence-shell electron pairs.

Rule 4. IN CYCLIC $(SN)_x$ SYSTEMS, THE HÜCKEL $4n + 2$ RULE APPLIES. A planar structure with delocalized π bonding is favored when n, the number of π electrons, is 2, 6, 10, or 14 (*2*). In such applications of simple Hückel MO theory, we treat $(SN)_x$ rings as if they were homonuclear: we ignore the electronegativity difference between sulfur and nitrogen. The effect of this approximation is to assign nonbonding character to some orbitals which otherwise would be weakly bonding (*3*). However, this approximate method is warranted by its simplicity, the agreement with valence bond formulation, and its success in structure correlation. In noncyclic chain systems π bonding is more localized.

Rule 5. LONE PAIR-LONE PAIR REPULSION BETWEEN DIRECTLY ATTACHED ATOMS IS STRUCTURALLY UNIMPORTANT. However, it can be significant between other atoms which are close together because of the molecular geometry.

(SN)ₓ Ring Systems

The simplest known $(SN)_x$ ring system is the compound S_2N_2. This molecule and its isoelectronic analog, $S_4{}^{2+}$, have square planar configurations (*4, 5, 6*). Because the ring has six π electrons, the $4n + 2$ Rule is obeyed, and the planar structure is stable. According to simple Hückel molecular orbital theory, the molecule has one pair of bonding π electrons, in agreement with valence bond theory:

Although the free radical $S_2N_2{}^+$ has been identified, the only available structural information is that the two nitrogen atoms are structurally equivalent and that the two sulfur atoms are structurally equivalent (*7*). The species is probably planar (or nearly planar) with the odd electron occupying a nonbonding π orbital.

The species $S_2N_2^{2+}$ is unknown at present. On the basis of its analogy to the P_4 molecule, one might predict a tetrahedral structure, however because of the tendency for bond angles at nitrogen atoms to be larger than those at second-row atoms, the acute bond angles in the tetrahedral structure may introduce so much instability as to make an oblong or rhomboidal planar structure, with two localized double bonds, stable:

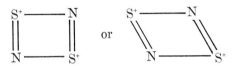

A six-membered $(SN)_3$ ring is in the molecule $S_3N_3Cl_3$, illustrated in Figure 1 and in the following valence-bond structure.

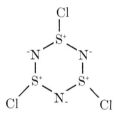

The tervalent sulfur atoms prohibit $p\pi$ bonding, and the ring has a non-planar conformation, analogous to the chair conformation of cyclohexane (8). However, the ring is nearly planar (the average distance of the nitrogen atoms from the plane of the sulfur atoms is only 0.2 A), and the

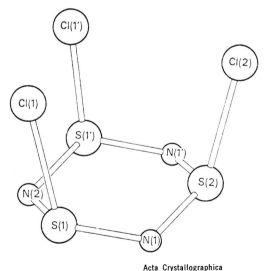

Acta Crystallographica

Figure 1.   The $S_3N_3Cl_3$ molecule (7)

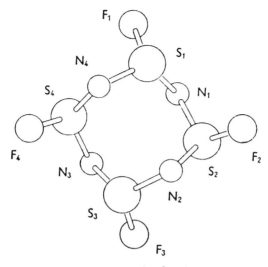

Acta Crystallographica

Figure 2.   The $S_4N_4F_4$ molecule (10)

$^+$S–N$^-$ bond distance (1.605 A) is shorter than the S–N distance in $S_4N_4H_4$ (1.65 A). Although these features suggest that $d\pi$–$p\pi$ bonding exists in the ring, the short $^+$S–N$^-$ bond distance may be a consequence of the polarity of the bond. The molecule is remarkable because all three chlorine atoms occupy apical rather than equatorial positions and because the nonbonding electron orbitals of each of the six pairs of adjacent sulfur and nitrogen atoms are essentially coplanar. One might have predicted that the chlorines would occupy equatorial positions, so that the nonbonding orbitals on adjacent atoms would be forced to lie in different planes. However, such a structure would cause the nonbonding electrons of the chlorine atoms to interact with those of the nitrogen atoms. Apparently the latter interaction is more repulsive than that between nonbonding electrons on adjacent ring atoms. This observation is used below in interpreting the structure of $S_4N_4F_4$.

The structure of $S_3N_3O_3Cl_3$ is almost the same as that of $S_3N_3Cl_3$ except that oxygen atoms occupy the positions of the sulfur lone pairs (9). In the isoelectronic compound $P_3N_3Cl_6$, the P_3N_3 ring is almost exactly planar. This near planarity may be caused by $d\pi$–$p\pi$ bonding and probably is favored in the P_3N_3 ring more than in the S_3N_3 ring because of the greater size of the phosphorus atom and the consequent relative ease of achieving an N–P–N bond angle near 120°.

The S_8 ring, if planar, would have 16 π electrons and consequently an equal number of antibonding and bonding π electrons. Thus, the nonplanar crown configuration, with bond angles of 108°, is stable. The sulfur imides correspond to S_8 rings in which 1, 2, 3, or 4 of the sulfur atoms have been replaced with the isoelectronic NH groups—with the restriction that the NH groups never occupy adjacent positions in the ring. Thus, as far as is known, S_7NH, $S_5N_3H_3$, $S_4N_4H_4$, and the isomers of $S_6H_2N_2$ have structures completely analogous to that of S_8 (*10*).

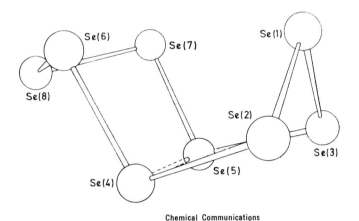

Chemical Communications

Figure 3.   The Se_8^{2+} molecule (12)

In $S_4N_4F_4$, the fluorine atoms are attached to the sulfur atoms, consequently the $(SN)_4$ ring has no $p\pi$ bonding and is highly puckered, as shown in Figure 2 (*11, 12*). The remarkable feature of this structure is that there are two S–N bond distances (1.66 and 1.54 A) which occur alternatingly in the ring. In Figure 2, the S_1–N_1 distance is 1.66 A and the N_1–S_2 distance is 1.54 A. By examination of Figure 2, it is seen that the S_1–N_1 and N_1–S_2 bonds are stereochemically different. The S_1–N_1 bond is longer probably because of repulsive interaction of the nonbonding electrons of fluorine atom F_1 with the nonbonding electrons of nitrogen atom N_1. The N_1–S_2 bond is shorter probably because there is no interaction between the nonbonding electrons of atoms F_2 and N_1. If atoms F_1 and F_3 were flipped to the upper side of the ring, there would be no resultant net advantage in terms of nonbonding electron interactions. In such a conformation, the S–N bonds would be expected to occur in the sequence s,s,l,l,s,s,l,l (s = short; l = long).

The Se_8^{2+} ion has the remarkable structure illustrated in Figure 3 (*13*). Although a planar monocyclic structure with 14 π electrons would fit the $4n + 2$ Rule, such a structure is ruled out by the requirement that the selenium bond angles be around 95°. The only planar structure even with bond angles as high as 105° is the following, in which two of the atoms are almost on top of each other.

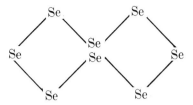

The observed structure of Figure 3 is derived from the normal crown conformation of Se_8 by the loss of two electrons, the transannular bridging from $Se(4)$ to $Se(5)$, and the ring flip of $Se(1)$. The species $S_4N_4^{2-}$, which is isoelectronic with Se_8^{2+}, never has been identified unequivocally (*14, 15*). The prediction of its structure is difficult. It might have a structure analogous to that of Se_8^{2+}, or it might have a planar π-bonded structure. A planar structure is not unreasonable for $S_4N_4^{2-}$ because of the ability of nitrogen atoms to have bond angles as high as $153°$ (*16, 17*).

Removal of two electrons from Se_8^{2+} to form the hypothetical Se_8^{4+} is expected to cause the flipping up of $Se(8)$ and the formation of a bridge from $Se(8)$ to $Se(1)$. This cage structure is analogous to that of the isoelectronic molecule, S_4N_4, illustrated in Figure 4 (*18*).

Removal of two electrons from S_4N_4 would yield the species $S_4N_4^{2+}$, which as yet has not been identified but which may be the stable sulfur–nitrogen cationic species which forms when S_4N_4 is dissolved in anhydrous sulfuric acid (*19*). This species would be isoelectronic with the cyclooctatetraenide ion $C_8H_8^{2-}$ and, because of adherence to the $4n + 2$ Rule,

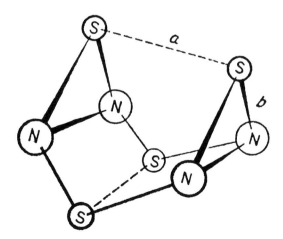

F. A. Cotton, G. Wilkenson, "Advanced Inorganic Chemistry," Interscience.

Figure 4. The S_4N_4 molecule (28)

would be expected to be a planar eight-membered ring of alternating sulfur and nitrogen atoms.

The $S_5N_5^+$ cation is a planar heart-shaped ring of alternating sulfur and nitrogen atoms (20):

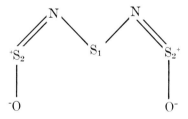

This ten-atom system has 14 π electrons, for which one predicts three π bonds.

Planar Sulfur–Nitrogen Chains

In this section we discuss chains of alternating sulfur and nitrogen atoms which are engaged in π bonding but which are not pseudoaromatic either because of an open chain structure or, for cyclic compounds, because of the presence of ring atoms which block π bonding.

The $S_3N_2O_2$ molecule is a planar chain of sulfur and nitrogen atoms terminated by oxygen atoms (21):

The indicated structure is probably the principal contributor to the resonance hybrid because it minimizes the separation of positive and negative formal charges and puts negative formal charges on the most electronegative atoms. As expected, the S_2–N distance (1.58 A) is shorter than the S_1–N distance (1.69 A).

The $S_3N_2Cl^+$ cation, pictured in Figure 5, has a planar S_3N_2 ring (22). Because of the inability of the tervalent sulfur atom to engage in π bonding, π bonding is restricted to the SNSN chain which is attached to the SCl group. Apparently the fact that the tervalent sulfur atom is coplanar with the SNSN chain is accidental. The bond angles in the ion are all

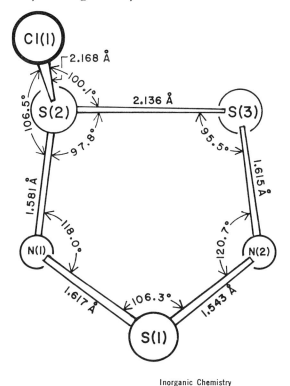

Inorganic Chemistry

Figure 5.   The S_3N_2Cl molecule (21)

normal. From the observed bond distances, we conclude that the π bond is largely localized as indicated in the following structure.

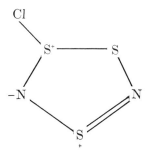

This structure is expected to be the principal contributor to the resonance hybrid because it involves the least separation of negative and positive formal charges.

The structure of $S_3N_3NPPh_3$ is shown in Figure 6 (23). Because the tervalent sulfur atom cannot engage in π bonding, the π bonding is restricted to the remaining NSNSN chain of atoms which are coplanar. It is not possible to form a completely planar six membered $(SN)_3$ ring with normal bond angles. However, by allowing the tervalent sulfur atom to flip out of the plane, the other five atoms can maintain planarity with normal bond angles at all six atoms. Because the π orbitals of the nitrogen atoms bonded to the tervalent sulfur atom are not oriented ideally, the π bonding would be expected to be strongest in the bonds to the middle nitrogen atom of the planar NSNSN chain.

The structure of S_4N_2 has not yet been determined unequivocally (24). One reasonable possibility is a structure like that of $S_3N_2Cl^+$ where the chlorine atom has been replaced by a sulfur atom. Another possibility is the six-membered ring structure below.

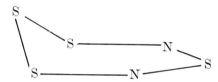

One would expect that π bonding would be restricted to the SNSNS chain of atoms and that these would be coplanar, with the middle sulfur atom of the SSS chain flipped out of the plane to allow normal bond angles at all the atoms. Nelson and Heal have shown recently that the physical properties of S_4N_2 are consistent with a six-membered ring structure with the nitrogen atoms separated by one sulfur atom (25). Although they point out that the data do not permit distinguishing between the possible boat, chair, and planar conformations, they do not even consider the conformation shown above.

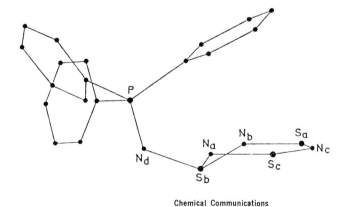

Chemical Communications

Figure 6.   The $S_3N_3NPPh_3$ molecule (22)

The $S_4N_3^+$ ion is a planar seven-membered ring, as shown below (*16, 17*).

The fact that the S–N bond distances are equal within experimental error indicates that there is delocalization of the π bonding in this ring (except in the S–S bond). The indicated structure is expected to be the main contributor to the resonance hybrid because it involves the least separation of negative and positive formal charges. The electronic absorption spectrum of the $S_4N_3^+$ cation has been interpreted in terms of $\pi \rightarrow \pi^*$ transitions (*26, 27*).

Literature Cited

(1) George, J. W., *Progr. Inorg. Chem.* (1960) **2**, 33.
(2) Streitwieser, A., "Molecular Orbital Theory for Organic Chemists," p. 256, Wiley, New York, 1961.
(3) Gleiter, R., *J. Chem. Soc. A* (1970) 3174.
(4) Warn, J. R. W., Chapman, D., *Spectrochim. Acta.* (1966) **22**, 1371.
(5) Patton, R. L., Raymond, K. N., *Inorg. Chem.* (1969) **8**, 2426.
(6) Stephens, P. J., *Chem. Commun.* (1969) 1496.
(7) Lipp, S. A., Chang, J. J., Jolly, W. L., *Inorg. Chem.* (1970) **9**, 1970.
(8) Wiegers, G. A., Vos, A., *Acta Crystallogr.* (1966) **20**, 192.
(9) Hazell, A. C., Wiegers, G. A., Vos, A., *Acta Crystallogr.* (1966) **20**, 186.
(10) Becke-Goehring, M., *Inorg. Macromol. Rev.* (1970) **1**, 17.
(11) Wiegers, G. A., Vos, A., *Acta Crystallogr.* (1961) **14**, 562.
(12) Wiegers, G. A., Vos, A., *Acta Crystallogr.* (1963) **16**, 152.
(13) McMullan, R. K., Prince, D. J., Corbett, J. D., *Chem. Commun.* (1969) 1438.
(14) Chapman, D., Massey, A. G., *Trans. Faraday Soc.* (1962) **58**, 1291.
(15) Meinzer, R. A., Pratt, D. W., Myers, R. J., to be published.
(16) Weiss, J., *Z. Anorg. Allg. Chem.* (1964) **333**, 314.
(17) Cordes, A. W., Kruh, R. F., Gordon, E. K., *Inorg. Chem.* (1965) **4**, 681.
(18) Lu, C.-S., Donohue, J., *J. Amer. Chem. Soc.* (1944) **66**, 818.
(19) Lipp, S. A., Jolly, W. L., *Inorg. Chem.* (1971) **10**, 33.
(20) Banister, A. J., Dainty, P. J., Hazell, A. C., Hazell, R. G., Lomborg, J. G., *Chem. Commun.* (1969) 1187.
(21) Weiss, J., *Z. Naturforsch.* (1961) **16b**, 477.
(22) Zalkin, A., Hopkins, T. E., Templeton, D. H., *Inorg. Chem.* (1966) **5**, 1767.
(23) Holt, E. M., Holt, S. L., *Chem. Commun.* (1970) 1704.
(24) Becke-Goehring, M., Fluck, E., "Developments in Inorganic Nitrogen Chemistry," p. 150, Vol. 1, C. B. Colburn, Ed., Elsevier, Amsterdam, 1966.
(25) Nelson, J., Heal, H. G., *J. Chem. Soc. (A)* (1971) 136.

(26) Johnson, D. A., Blyholder, G. D., Cordes, A. W., *Inorg. Chem.* (1965) **4,** 1790.
(27) Friedman, P., *Inorg. Chem.* (1969) **8,** 692.
(28) Cotton, F. A., Wilkinson, G., "Advanced Inorganic Chemistry," 2nd ed., Interscience, New York, 1966.

RECEIVED March 5, 1971.

7

The Influence of Pressure and Temperature on the Structure of the Sulfur Molecule and Related Structural Changes in Other Group VI A Elements: From Dimer through Ring and Chain to Metal

GARY C. VEZZOLI and ROBERT J. ZETO

The Institute for Exploratory Research, U. S. Army Electronics Command, Fort Monmouth, N. J. 07703

A pressure and temperature induced structural progression exists in sulfur from paramagnetic dimer through high resistance octameric puckered rings and lower resistance helical chains to a metallic state. This structural development in sulfur from the oxygen-like (dimer) through the selenium- and tellurium-like phase (chain) to a polonium-like form (metal) is related to some degree to structural changes in other group VI A elements. It is possible to develop a general diagram for group VI A which emphasizes the chain modification, the maxima in the melting curves, and the transition to the metallic state. Similarities in the pressure–temperature relationships for these elements in the liquid state are also evident, suggesting the dissociation of the liquid's chain structure induced by temperature and pressure.

Elemental sulfur belongs to group VI A of the periodic table of which oxygen, selenium, tellurium, and polonium are also members. At ambient conditions sulfur consists of octameric puckered rings (S_8) (*1, 2*) stacked in an orthorhombic lattice and is an insulator of high resistivity which is soluble in CS_2.

*The Pressure–Temperature (p–T) Relationships in Sulfur
in the Solid State*

A crystalline phase of sulfur has been synthesized by workers (3, 4, 5, 6, 7) at pressures above 21 kb at temperatures in excess of 210°– 240°C. This phase is insoluble in CS_2. Lind and Geller (8) showed that this high pressure phase, designated fibrous sulfur, is monoclinic (P2) and consists of right and left handed helices. This helical structure resembles slightly the helicoidal zigzag chain structures of selenium and tellurium at ambient conditions. Fibrous sulfur has several properties characteristic of a modest semiconductor such as a steep negative temperature coefficient of resistance and a photoelectric effect. It shows also an EPR absorption peak and does not revert at ambient conditions (for up to 4 years) to ordinary octameric sulfur (6, 7).

Indications have been reported for metallic conduction in sulfur at high temperature at ultra high pressures ranging from 87 to 230 kb (9, 10, 11). The metallic conduction is believed to arise from a transition to a metallic structure, perhaps similar to that of polonium. However, there is some question as to whether this transition took place experimentally in the solid state (9, 12) as a result of the uncertainty in the location of the liquidus at ultra high pressures. The metallic sulfur state is characterized by a conductivity which decreases with increasing pressure and has electrical properties different from typical metals (11).

When sulfur is quenched at reduced pressure from the temperature at which the vapor species is predominantly diatomic S_2 to liquid nitrogen temperature, a diatomic solid-state modification is formed with a structure similar to that of oxygen, being paramagnetic with two unpaired electrons, and this structure can be retained up to −80°C (13, 14, 15, 16, 17).

Thus, by changing the intensive variables pressure and temperature, the structure and properties of sulfur are transformed to span the entire range of group VI A, from paramagnetic dimer through insulating rings and semiconducting chains to the metallic state or from oxygen-like through selenium- and tellurium-like structures to a metallic or a quasi-metallic form. The $p–T$ ranges in which the various solid state modifications of sulfur are formed are shown schematically in Figure 1.

The interesting polymorphism observed in sulfur is consistent with previous predictions. Bridgman (19) named sulfur as a possible candidate for new irreversibly created high pressure and high temperature modifications. Von Hippel (1) showed that the structures of selenium and tellurium are interrelated to the structure of polonium and can be developed by displacing the atoms of the octahedral planes or sliding the (111) planes in the polonium structure. He reported that at ambient conditions each sulfur atom has primary coordination number 2 because more energy is gained in forming two single bonds as in a ring (102 kcal/mole) (*20*) than one double (molecular orbital) bond (<88 kcal/mole) (*21*). The formation of the chain-type sulfur polymorph was predicted by Deaton and Blum (*22*) to precede any transition to the metallic state.

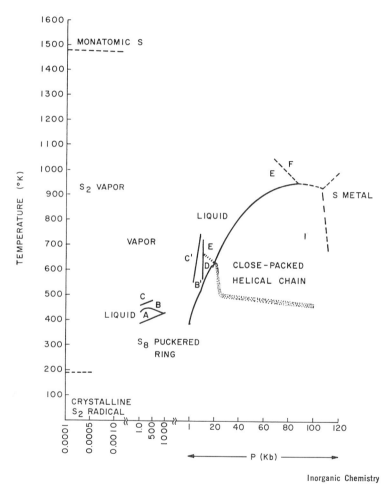

Inorganic Chemistry

Figure 1.   Proposed p-T *diagram for sulfur*

Heavy line indicates liquidus. Dashed boundaries at high p–T are inferred from the similarity between the phase equilibria of sulfur and that of tellurium. Dotted zone represents minimal experimental p–T conditions for the formation of fibrous sulfur (6). (For liquid state see Ref. 18.)

A = *short chains plus rings*	B = *polymer plus rings*
C = *dissociating polymer*	B' = *plastic product*
C' = *brittle product*	D,E = *viscoelastic rubbery product*
Hatched zone = *product resembling B'*	F = *dense dissociating liquid*

Analogous p–T Relationships in Selenium, Tellurium,
and Polonium in the Solid State

From evidence of diatomic selenium and tellurium in the vapor state
and of octameric puckered rings in monoclinic selenium, as well as evidence of metallic tellurium at high pressure (23), it is suspected that in
principle (if sluggish kinetics can be overcome) pressure and temperature
may induce transformations in other group VI A elements comparable
with those observed in sulfur.

A phase transition in selenium has been observed at room temperature at ca. 55 kb. However, little is known about the nature of this transition. Tellurium is known to have at least two and possibly four high
pressure phases (24). The semiconductor to metallic transformation has
been well established at ca. 40 kb at room temperature (4, 5); however,
the structure of the high pressure metallic phase has not been elucidated
unambiguously. The high pressure x-ray diffraction pattern of this phase
has been reported by Jamieson and McWhan (25) and by Kabalkina,
Vereschagin, and Shulenin (26) and can be indexed as a body-centered
orthorhombic lattice (27). The calculated lattice parameters based on
this structure give rise to a theoretical x-ray density which agrees with
the value that can be derived from the volume vs. pressure experiments
on tellurium by Bridgman. The semiconductor to metallic $p–T$ phase
boundary intersects the liquidus at the cusp (beyond the liquidus maximum) at about 29 kb at 450°C (28, 29). At a pressure of 70 kb at room
temperature Jamieson and McWhan (25) have observed, again by high
pressure x-ray diffraction, a transformation in tellurium to the β-polonium
structure.

Two crystalline modifications of polonium exist at atmospheric pressure—the low temperature α-cubic form and the high temperature β-rhom-
bohedral form (30). The α to β transition occurs between 75° and
100°C; however, the β to α transition is observed at about 10°C; thus
the equilibrium temperature for the transition has not been established.
Polonium has been difficult to study because of its radioactive nature
and its decay into lead.

Maxima in the Melting Curves and Relationships in the
Liquid State for Group VI A

Selenium and tellurium have the A8 helicoidal zigzag chain structure
at ambient conditions (1). Both show maxima in their melting curves
at about 50 kb at 963°K for selenium (32) and at an average $p–T$ of
about 12 kb at 743°K (22, 29, 33, 34, 35) for tellurium. These maxima
are suspected to be related to the maximum in the melting curve of sulfur
at 86 kb at 953°K (12) because at these conditions the sulfur polymorph

which is expected to be in equilibrium with the liquid (*36*) is helical fibrous sulfur, which although a much more complex structure than hexagonal selenium, still is based on chain molecules.

Similar *p–T* relationships exist also in the liquid state in sulfur, selenium, and tellurium. Gee (*37*) has studied the polymerization of liquid sulfur extensively. At high temperatures at atmospheric pressure all three elements show a thermally induced dissociation of the liquid chain structure with a gradual change toward a lower resistance liquid (*38, 39, 40, 41, 42*). In the liquid state *p–T* reaction boundaries have been observed in sulfur (*18, 43*) and tellurium (*31, 44*). For sulfur these are consistent with the high temperature atmospheric–pressure data (*39, 40*). For tellurium, however, the low pressure extrapolation of the liquid boundaries must be nonlinear to achieve consistency. The *p–T* reaction boundaries in the liquid for sulfur are plotted in Figure 1.

In tellurium a boundary in the liquid state has been indicated by Deaton and Blum (*22*) which extends from the temperature at which tellurium begins to show metallic conduction at 1 atm (943°K) (*41*) to the neighborhood of the melting curve maximum (*22*) and is postulated to represent chain dissociation. Above this boundary metallic type behavior appears to dominate the conductivity (*22*). Controversy over this boundary is apparent also as Stishov (*44*) reports only semiconducting behavior, rather than metallic, in the liquid tellurium high pressure fields and states that the boundary intersects the liquidus beyond the maximum.

Another *p–T* boundary in liquid tellurium appears to extend from the temperature (848°K) at 1 atm where the Hall coefficient changes sign from positive to negative (*41*) down to the liquidus, perhaps intersecting the latter at a discontinuous slope change in the melting curve at 3.6 kb at 736°K (*34*). This boundary is believed to be associated with the dissociation of the liquid chain structure giving rise to fewer electron vacancies (holes) and more negative carriers, thus explaining the change in Hall coefficient sign. Since both liquid boundaries in tellurium appear to involve chain dissociation and are sloped negatively, pressure (as well as temperature) inhibits the preservation of the chain structure in the liquid state (*22, 41*). These dissociation boundaries are not true phase boundaries and probably have a statistical nature which is probably true of other *p–T* boundaries in the liquid state in group VI A elements. The fact that the chain structure in the liquid state is dissociating with increasing pressure, whereas in the solid state it is compacting with pressure, may explain why the density of the liquid becomes equal to that of the solid with which it is in equilibrium at the melting curve maximum and greater than the density of the solid over the negatively sloped portion of the melting curve beyond the maximum.

Proposed Phase Equilibria for Sulfur

From observations of metallic conduction in sulfur at ultra high pressures (9, 10, 11) and from correlation between the phase equilibria of sulfur and that of tellurium, we suspect that in the crystalline state a semiconductor to metallic boundary exists in sulfur which intersects the liquidus at a cusp beyond the maximum at 86 kb at 953°K. We also suggest that in the liquid state in sulfur a chain dissociation boundary similar to that observed in liquid tellurium probably exists and intersects the liquidus near the maximum at 86 kb. Although chain dissociation in the liquid state has not been studied in detail at high pressure, a boundary of the above type in sulfur is thought to represent the dissociation of close packed liquid E in Figure 1 or an extensive state of depolymerization. The boundary may be related to the formation of diatomic or monatomic sulfur which are shown at high temperature and low pressure in Figure 1, or to a possible C'–E boundary.

General Phase Equilibria in Group VI A

From observations of rings, chains, metallic states, and melting curve maxima in group VI A elements, the following general trend on increasing pressure and temperature is proposed: dimeric paramagnetic radical → insulating octameric puckered rings → semiconducting helical chains → metallic. All of these structural forms have not been observed in each individual member of group VI A. Size and energy criteria may prohibit the existence of some structural forms of certain group members, and some structural modifications may be stable at p–T conditions beyond the capability of present apparatus. Only in sulfur has the entire structure and property spectrum from dimer through rings and chains to metal been reported.

Deviations from the general trend exist within this group. Fibrous sulfur does not have the A8 structure of selenium and tellurium, and the helical sulfur molecule ($^{10}S_3$) is different from the helical selenium molecule ($^3S_1$). Also tellurium rings have not been observed, and metallic sulfur does not have properties similar to metallic tellurium or polonium. Thus, any generalized phase diagram will have serious limitations.

The sulfur p–T diagram shown in Figure 1 contains several features of the diagrams of selenium, tellurium, and polonium—*i.e.*, chain field (Se and Te), maximum in the melting curve (Se and Te), and metallic phase (Te and Po). The structural progression and phase equilibria among members of group VI A are correlated and depicted schematically in Figure 2. The figure can be viewed as the sulfur phase diagram with translations of the p–T coordinate origin relative to the maxima in the melting curves of selenium and tellurium and with appropriate changes in the pressure and temperature scales. Not enough is known about specific structural changes in these elements to specify the exact location of the appropriate *p–T* axes on the general diagram. The pressure axes (corresponding to the isotherm zero absolute temperature) are situated so as to demonstrate continuity of the chain structure through sulfur, selenium, and tellurium and the transition to the metallic state. The extension of the pressure axis to the left of the origin represents pressure less than atmospheric or volume expansion.

In Figure 2 the temperature axes (representing the atmospheric pressure isobar) are positioned schematically in proportion to the number of kilobars between atmospheric pressure and the pressure corresponding to the melting curve maxima in sulfur, selenium, and tellurium. Thus the melting curve maxima are superimposed on the general diagram, and these maxima may represent fundamentally much more than just a virtual origin because a maximum in the liquidus specifies a phase change

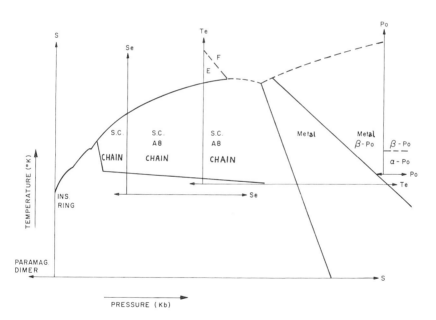

Figure 2.   Proposed general p-T *diagram for group VI A elements*

The general diagram is derived from the sulfur diagram of Figure 1 by translating the origin of the p–T axes relative to the maximum in the melting curve and using arbitrary p–T scales. The general or approximate location of the new temperature axes (corresponding to 1 atm pressure) relative to the melting curve maximum occur for S, Se, and Te in the same succession as do these elements in group VI A in the periodic chart.

E–F = chain dissociation boundary   *Ins. = insulating properties*
S.C. = semiconducting properties   *Met. = metallic properties*
Paramag. = paramagnetic (unpaired electron spins)
– – – = inferred or suggested boundaries

at zero volume change. This suggests the development of a more funda-mental and unified group VI A diagram using reduced parameters.

Geller (45) observed that at high pressure a solid solution with selenium–sulfur chemical bonds could be formed and that also at high pressure unbonded spiraling helices of fibrous sulfur and tellurium could be synthesized as Te_7S_{10} (46). This formation of Te_7S_{10} occurs because of almost identical van der Waals radii of the constituents and because a seven-atom increment of tellurium helices has the same length as a ten-atom increment of a sulfur helix. One criteria for solid solutions (such as selenium–sulfur) to be formed is that structures must be generally similar. However, a maximum sulfur content near $S_{.56}Se_{.44}$ has been reported by Geller (46), despite the fact that sulfur forms a complete solid solution range with octaselenium.

The general diagram in Figure 2 may help explain why no tellurium rings have been observed experimentally—namely because extremely low pressures or high degrees of expansion are required to form the ring struc-ture at accessible temperatures. The kinetics for a chain to ring reaction completely in the solid state in tellurium would be extremely slow because of the low temperatures involved and because chain structures, once formed, are interlinked by so much secondary bonding that it is almost impossible for them to revert to ring configurations (1).

Polonium's position on the general diagram is uncertain. The effect of pressure on the α-to-β polonium transition is not known, but the densi-ties of the α and β phases are the same at atmospheric pressure, about 9.4 ± 0.5 grams/cc (30). Hence, from the Clapeyron equation we would expect that the p–T boundary should be almost horizontal, perhaps with slight negative slope to indicate as Bridgman suggests that high pressure usually favors the less symmetric phase, in this case the rhombohedral β form. From the evidence available only the highest pressure portion of the general diagram in Figure 2 can be applied tentatively to polonium.

At high pressure and low temperature tellurium, and perhaps even selenium and sulfur, may undergo a transition to the α-polonium struc-ture. This probably is not observed experimentally without great diffi-culty because of the probable extreme p–T conditions associated with formation of the phase and the expectation that it could not be recovered to be analyzed at ambient conditions thus requiring a sophisticated high pressure low temperature x-ray camera.

A further correlation between tellurium and sulfur should be men-tioned. In the solid state tellurium changes from negative to positive Hall coefficient at 503°K (41) at atmospheric pressure and at 519°K at 2 kb (47). At several pressures to 13 kb a resistance discontinuity in tellurium has been observed in solid media apparatus (48) also at 503°K. In sulfur the approximate temperature and slope of the boundary above which the close-packed chain structure can be quenched (the dotted almost-horizontal boundary in Figure 1 appears to be almost identical to the above "boundary" in tellurium). Whether this correspondence is coincidental or fundamental is not clear; however, a positive Hall coefficient indicates predominant hole conduction which is associated with a "tight" or close-packed chain-type structure in which electrons are used for bonding and electron vacancies for conduction.

The pressure and temperature studies of sulfur, selenium, and tellurium conducted by the workers cited herein help explain what first appears unusual when glancing at group VI A on the periodic chart— why group VI A elements have such varying properties and structures at ambient conditions, yet are in the same family having the same valence bond s^2p^4 configuration. The explanation seems to lie in the observations that the covalent van der Waals bonding in these elements are quite pressure and temperature sensitive so that changes in the strong covalent bonds give rise to the major structural changes (dimer → ring → chain → metal) whereas alterations in the weaker van der Waals bonds give rise to some of the property changes (solubility and viscosity); the end result is that pressure tends to converge the properties and probably the structures of these elements to the most metallic form.

Figure 2 depicts schematically the similarities between the p–T relationships in group VI A elements and is not intended to be interpreted as a strict generalized phase diagram for group VI A. Only similarities in structure and properties can be inferred from a diagram of this type, and generalizations are limited severely because of the number of deviations which exist and because not enough detailed quantitative information is known. The general trend through this group is nonetheless important.

The structural progression in group VI A elements is consistent with the expectation that ultra high pressure should transform all structures to the low volume metallic state, and indeed the advent of super pressure technology has shown that no longer can a material be designated insulator, semiconductor, metal, or supermetal except by specifying rigorously the conditions of intensive variables to which the material is and has been subjected.

Literature Cited

(1) Von Hippel, A., *J. Chem. Phys.* (1948) **16**, 372.
(2) Cotton, F., Wilkinson, G., "Advanced Inorganic Chemistry," pp. 522–526, Interscience, New York, 1966.
(3) Geller, S., *Science* (1966) **152**, 644.
(4) Sclar, C., Carrison, L., Gager, W., Stewart, O., *J. Phys. Chem. Solids* (1966) **27**, 1339.
(5) Vezzoli, G. C., Dachille, F., Roy, R., *Science* (1969) **166**, 218.

(6) Vezzoli, G. C., Dachille, F., *Inorg. Chem.* (1970) **9**, 1973.
(7) Vezzoli, G. C., Zeto, R. J., *Inorg. Chem.* (1970) **9**, 2478.
(8) Lind, M., Geller, S., *J. Chem. Phys.* (1969) **51**, 348.
(9) Berger, J., Joigneau, S., Bottet, G., *Compt. Rend. Acad. Sci.* (1960) **250**, 4331.
(10) David, H., Hamann, S., *J. Chem. Phys.* (1958) **28**, 1006.
(11) Joigneau, S., Thouvenin, J., *C. R. Acad. Sci.* (1958) **246**, 3422.
(12) Susse, C., Epain, R., Vodar, B., *J. Chem. Phys. France* (1966) **63**, 1502.
(13) Berkowitz, J., Marquart, J., *J. Chem. Phys.* (1963) **39**, 275.
(14) Cotton, F., Wilkinson, G., "Advanced Inorganic Chemistry," pp. 524–525, Interscience, New York, 1966.
(15) Meyer, B., *J. Chem. Phys.* (1962) **37**, 1577.
(16) Radford, H., Rice, F. O., *J. Chem. Phys.* (1960) **33**, 774.
(17) Rice, F. O., Sparrow, C., *J. Amer. Chem. Soc.* (1953) **75**, 848.
(18) Vezzoli, G. C., Dachille, F., Roy, R., *J. Polymer Sci.* (1969) **7 (Pt. A-1)**, 1557.
(19) Bridgman, P. W., "Solids Under Pressure," p. 8, W. Paul and D. Warshauer, Eds., McGraw-Hill, New York, 1963.
(20) Pauling, L., "The Nature of the Chemical Bond," p. 85, Cornell University Press, Ithaca, N. Y., 1960.
(21) Samuel, R., *Rev. Mod. Phys.* (1946) **18**, 114.
(22) Deaton, B., Blum, F., *Phys. Rev.* (1965) **137**, A 1131.
(23) Bridgman, P. W., *Proc. Am. Acad. Arts Sci.* (1952) **81**, 169.
(24) Stishov, S., Tikhomirova, N., *Zh. Eksp. Teor. Fiz.* (1965) **49**, 618; *JETP* (1966) **22**, 429.
(25) Jamieson, J., McWhan, D., *J. Chem. Phys.* (1965) **43**, 1149.
(26) Kabalkina, S., Vereshchagin, L., Shulenin, B., *Zh. Eksp. Teor. Fiz.* (1963) **45**, 2073; *JETP* (1964) **18**, 1422.
(27) Vezzoli, G. C., Z. *Kryst*, in press.
(28) Blum, F., Deaton, B., *Phys. Rev.* (1965) **137**, A 1410.
(29) Klement, W., Jr., Cohen, L., Kennedy, G., *J. Phys. Chem. Solids* (1966) **27**, 171; also Kennedy, G., Newton, R., in Ref. *19*, p. 163.
(30) Beamer, W., Maxwell, C., *J. Chem. Phys.* (1949) **17**, 1293; Maxwell, C., *J. Chem. Phys.* (1949) **17**, 1288.
(31) Vezzoli, G. C., Zeto, R. J., *Inorg. Chem.*, in preparation.
(32) Paukov, I., Tonkov, E., Mirinskiy, D., *J. Phys. Chem. Moscow* (1967) **8**, 995.
(33) Chaney, P., Babb, S., *J. Chem. Phys.* (1965) **43**, 1071.
(34) Stishov, S., Tikhomirova, N., Tonkov, E., *Zh. Eksp. Teor. Fiz.* (1966) **4**, 161.
(35) Tikhomirova, N., Stishov, S., *Zh. Eksp. Teor. Fiz.* (1962) **43**, 2321; *JETP* (1963) **16**, 1639.
(36) Vezzoli, G. C., Dachille, F., Roy, R., *Inorg. Chem.* (1969) **8**, 2658.
(37) Gee, G., *Trans. Faraday Soc.* (1952) **48**, 515.
(38) Vezzoli, G. C., *ECOM Tech. Rept.*, in preparation.
(39) Vezzoli, G. C., *J. Polymer Sci.* (1970) **8 (Pt. A-1)**, 1587.
(40) Vezzoli, G. C., *J. Amer. Ceram. Soc.*, in press.
(41) Epstein, A., Fritzsche, H., Lark-Horovitz, K., *Phys. Rev.* (1957) **107**, 412.
(42) Johnson, V. A., *Phys. Rev.* (1955) **98**, A 1567.
(43) Vezzoli, G. C., Ph.D. Dissertation, The Pennsylvania State University, University Park, Pa. (March 1969).
(44) Stishov, S., *Zh. Eksp. Teor. Fiz.* (1967) **52**, 1196; *JETP* (1967) **25**, 795.
(45) Geller, S., Lind, M., *J. Chem. Phys.* (1970) **52**, 3782.
(46) Geller, S., *Science* (1968) **161**, 290.
(47) Nussbaum, A., Myers, J., Long, D., *Phys. Rev. Lett.* (1959) **2**, 6.
(48) Vezzoli, G. C., Zeto, R. J., *J. Appl. Phys.*, in preparation.
RECEIVED August 20, 1971.

The Reaction of Mercaptans with Liquid Sulfur

H. J. LANGER[a] and J. B. HYNE

Alberta Sulphur Research Ltd., University of Calgary, Calgary, Canada

The kinetics of the reactions of various substituted thiophenols with liquid sulfur have been investigated, and intermediates of the general form RS_xH and HS_xH have been identified. The studies suggest that the reaction is usually free radical involving an initiation period in which a steady state concentration of sulfur radical species is established. The reaction appears to be second order in mercaptan and third order in sulfur. These findings have been interpreted using a complex sulfur radical species involving several S_8 molecules.

Recent studies of the reaction of hydrogen sulfide with liquid sulfur (1, 2, 3) have indicated that the reaction is probably free radical in nature and yields a variety of hydrogen polysulfides or sulfanes. No comparable study of mercaptans with liquid sulfur has been reported, however, despite interest in this system in the vulcanization, petrochemical, and geochemical fields. The RSH–sulfur reaction offers advantages over the H₂S–S system since the effect of varying the R group in the mercaptan is used as an additional probe in elucidating the reaction characteristics. The results of kinetic and product analysis studies of the reaction of a range of p-substituted aromatic mercaptans (R = X–C₆H₄–) with liquid sulfur are presented. Experimental details will be published in a separate more extensive paper (4).

Kinetic Studies

Order of Reaction. Typical second order rate plots for the reaction of the parent thiophenol with liquid sulfur at 130°C are shown in Figure

[a] Present address: Ashland Oil Inc., Columbus, Ohio.

113

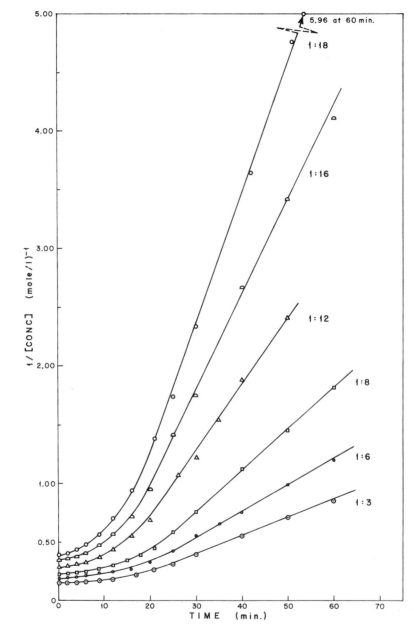

Figure 1. Second order (with respect to thiol) rate plots for the reaction of thiophenol with sulfur (130°C). Ratios shown on each line are thiophenol:sulfur.

1. The decrease in concentration of thiophenol was followed by aliquot sampling of the reaction mixture and determination by NMR of the relative strength of the mercaptan proton signal. The concentration/time

data obtained were examined to determine the order of reaction with respect to thiophenol. A first order analysis was shown to be unsatisfactory. The second order plots shown in Figure 1 are characterized by an initial curved section followed by a linear relationship between reciprocal concentration and time. The various plots represent different relative concentrations of thiol:sulfur. A range of relative concentrations from 1:18 thiol to sulfur to 1:3 thiol to sulfur were examined. By virtue of the excess sulfur in each run, the concentration of sulfur is assumed to remain relatively constant, at least within the limits of accuracy of the rate constants obtained. The linearity of the reciprocal concentration–time plots supports the pseudo-second order rate expression.

$$- \frac{d[\phi SH]}{dt} = k_{obs}[\phi SH]^2 \tag{1}$$

The k_{obs} values obtained from such a second order analysis are shown as a function of relative sulfur concentration in Table I. The observed dependence of k_{obs} on sulfur concentration indicates that the rate expression shown in Equation 1 is inadequate and that a concentration term in sulfur is required. Assuming an unknown order of reaction b with respect to sulfur, an expanded rate expression is obtained.

$$- \frac{d[\phi SH]}{dt} = k_{ov}[\phi SH]^2[S]^b \tag{2}$$

Thus
$$k_{obs} = k_{ov}[S]^b \tag{3}$$

and
$$log\ k_{obs} = \log k_{ov} + b\ log\ [S] \tag{4}$$

Hence, a plot of $log\ k_{obs}$ *vs.* $log\ [S]$ should yield a straight line of slope b, the order of reaction with respect to sulfur. Such a plot is shown in Figure 2 with lines having slopes corresponding to b values of 2, 3, and 4 shown for comparison. The value of $b = 3$ accommodates best the experi-

Table I.   Thiophenol–Sulfur Reaction[a]

Molar Ratio $\phi SH{:}S$	Initial $[\phi SH]$ mole l^{-1}	Initial $[S]$ mole l^{-1}	$k_{obs} \times 10^3$ $l\ mole^{-1}\ sec^{-1}$
1:18.8	2.540	47.70	1.84
1:16	2.850	45.64	1.35
1:12	3.465	41.59	0.938
1:8	4.411	35.30	0.585
1:6	5.110	30.67	0.375

[a] Second order (with respect to ϕ–SH) rate constants at 130°C as a function of relative sulfur concentration.

mental data. Thus, the overall rate expression suggested by the rate data for the reaction of thiophenol with liquid sulfur is

$$-\frac{d[\phi\text{SH}]}{dt} = k_{ov}[\phi\text{SH}]^2[\text{S}]^3 \qquad (5)$$

Induction Period and Effect of Radical Scavengers. The pseudo-second order plots in Figure 1 indicate that the reaction is characterized by an induction period—the initial nonlinear portion preceeding the pseudo-second order linear behavior. This characteristic of the rate plots was observed in all reactions studied involving variation of concentration, temperature, and ring substitution, except for the cases of p-amino and p-nitrothiophenol. It is suspected that a different reaction mechanism is applicable to these two thiols. Such induction periods are characteristic of radical initiated reactions and represent the initial reaction period required to establish a steady state condition with respect to the reaction propagating species in the system. With this hint of radical involvement in the reaction pathway the effect of added radical scavengers was investigated. Both phenol and hydroquinone are shown in Figure 3 to have an effect on the extent of the induction period and the value of k_{obs} (note the reduction of the slope of the pseudo-second order plots compared with the reference case with no inhibitor added). The existence of the induction period and the inhibiting effect of known radical scavengers support the suggestion of a radical mechanism.

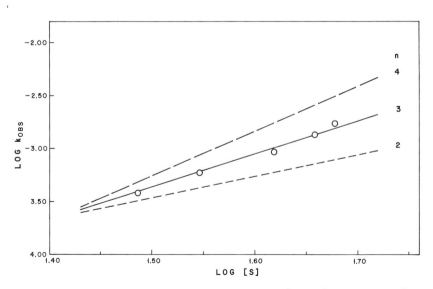

Figure 2.   Determination of reaction order with respect to sulfur

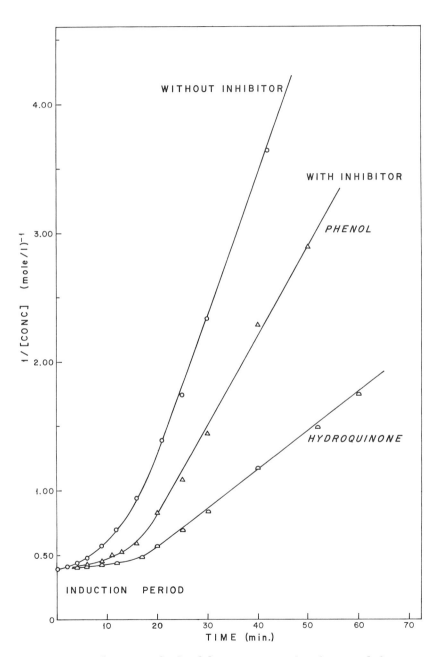

Figure 3. Effect of radical inhibitors on second order rate behavior of thiophenol/sulfur reaction (130°C)

Temperature Effects. The effect of varying temperature on the rate of the parent thiophenol–sulfur reaction was examined over the range 130°–160°C. Apart from limitations caused by the rapidly increasing reaction rate, the constraints of the melting point of sulfur and the equilibrium polymerization temperature with concomitant viscosity increase were responsible for the relatively narrow, kinetically accessible temperature range. Nonetheless a good Arrhenius type plot of the observed pseudo-second order rate constant, $log\ k_{obs}$, vs. reciprocal temperature was obtained (Figure 4), yielding values for the apparent enthalpy of activation, $\Delta H^* = 31.6 \pm 1.2$ kcal, and entropy, $\Delta S^* = 7.0 \pm 2.7$ eu.

Substituent Effects. The effect on the rate of a variety of electron-donating and -withdrawing para substituents on thiophenol were examined. Pseudo-second order analysis similar to those shown in Figure 1 were carried out but for only one thiol:sulfur concentration ratio. The rate data obtained under specified conditions are shown in Table II. The thiols have been listed in order of increasing Hammett σ value of the substituents—i.e., most effective electron donor to most effective withdrawer. Examination of these data shows that there is a marked substituent effect on the reaction. The electron-donating para substituents, methoxy and methyl, increased the rate while the electron-withdrawing groups, fluoro, chloro, and bromo, decreased the rate. Two substituents, p-amino and p-nitro, are anomalous in that the reaction rates do not obey the Hammett

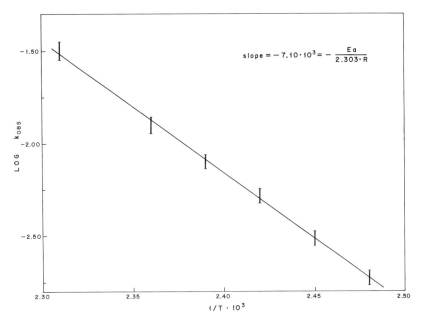

Figure 4.   Arrhenius plot for thiophenol/sulfur reaction

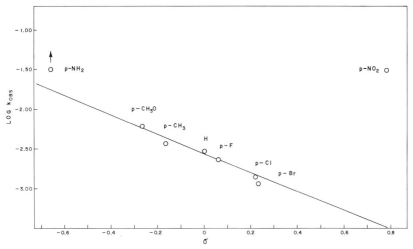

Figure 5.   Effect of substituents on thiophenol/sulfur reaction—Hammett type plot

type correlation. The *p*-amino compound reacted too rapidly with sulfur to allow quantitative determination of the rate constant, and the reactivity of the *p*-nitro compound is comparable with that of the *p*-amino compound rather than being slowest of all the substituted thiols examined as expected from the σ value. The linear free energy relation is illustrated in Figure 5. Apart from the *p*-amino and *p*-nitro substituents, a good linear correlation between the reactivity of the thiols toward liquid elemental sulfur and the electron-releasing or electron-withdrawing ability of each substituent is measured by its σ value is obtained. From the slope of this line the reaction constant, ρ, was -1.2.

Table II.   Rates of Reaction of Para-Substituted Thiophenols with Liquid Elemental Sulfur at 130°C with $[ArSH]_o/[S] = 1{:}18.8$

Z in $Z–C_6H_4–SH$	$[ArSH]$[b] $mole\ kg^{-1}$	$[S]$[b] $mole\ kg^{-1}$	$k_{obs} \times 10^3$ $kg\ mole^{-1}\ sec^{-1}$	σ Value of Z[a] Substituent
p-NH$_2$	1.374	25.85	>31.5	−0.66
p-CH$_3$O	1.346	25.30	6.3	−0.268
p-CH$_3$	1.376	25.86	3.7	−0.170
H	1.405	26.40	3.0	0.000
p-F	1.368	25.72	2.3	+0.062
p-Cl	1.338	25.15	1.4	+0.227
p-Br	1.263	24.75	1.1	+0.232
p-NO$_2$	1.320	24.80	31.5	+0.778

[a] Hammett σ values from Ref. 5.
[b] Initial concentrations.

From the Hammett plot we see that the *p*-amino and *p*-nitro substituted thiophenols do not fit this linear relationship. This fact suggests that these two thiols react by a different mechanism from that for the other thiols.

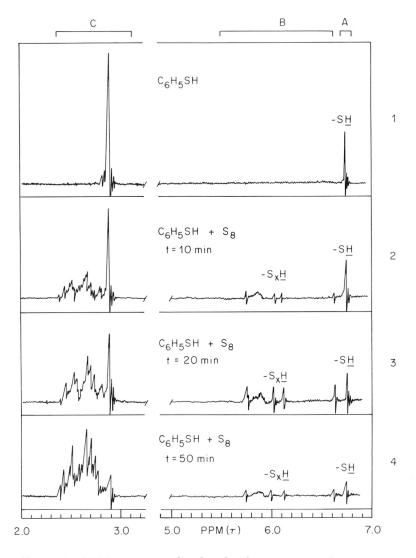

Figure 6.   NMR spectra for thiophenol/sulfur reaction with increasing (1–4) reaction time. Region A—SH signal decreases with time. Region B—polysulfanes appear and subsequently decrease. Region C—decrease in C_6H_5 signal and increase in product $(C_6H_5)_2S_x$.

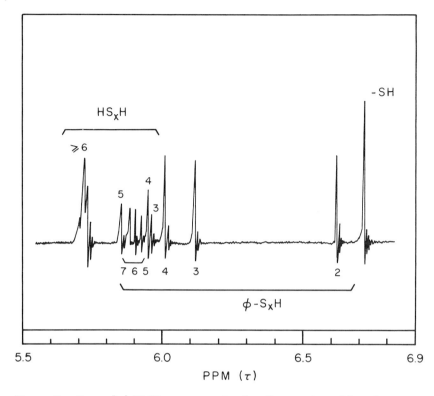

Figure 7.   Expanded NMR spectrum of polysulfane region with assignments. Numbers indicate value of x.

Intermediates

The course of the overall reaction as shown in Equation 6

$$2 \, ArSH + 1/8 \, S_8 \rightarrow ArS_xAr + H_2S \quad (x \text{ predominantly } 2) \qquad (6)$$

was investigated by NMR under the same conditions and procedures used in the kinetic studies. In the kinetic studies the primary interest was the rate of disappearance of the thiophenol–SH signal, but in addition to observing the diminution of the –SH signal, the appearance and disappearance of other NMR signals also characterized the NMR study. By analyzing the NMR spectra of the reaction mixture at various times, it was found that while the absorption of the sulfhydryl peak decreased, new peaks downfield of –SH appeared and then disappeared during the reaction, indicating that intermediates were formed. An illustration of this behavior is presented in Figure 6 where NMR spectra, taken from the reaction mixture of the reaction between benzenethiol and liquid elemental sulfur as a function of time, are shown. Figure 7 is an expanded

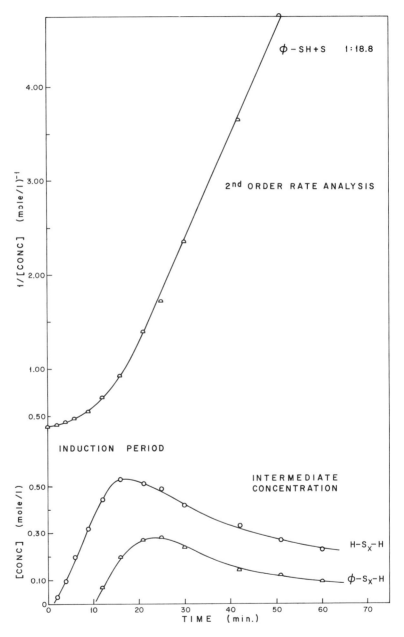

Figure 8.   Relationship between induction period for thiophenol/sulfur reaction and concentration variation of intermediates

section of the less detailed reaction spectrum shown in Figure 6. The NMR signals resulting from HS_xH species in the spectrum shown in Figure 6 were assigned according to Hyne *et al.* (2). The formation of these sulfanes was not unexpected since it has been shown that HS_xH species are formed by the reaction of H_2S, a product of the reaction, and molten sulfur (6).

In addition to the HS_xH NMR signals, however, other transient peaks appeared between the absorption of HS_xH and –SH. Since the polysulfanes, HS_xH, are products of the reaction of HSH with liquid elemental sulfur, the formation of sulfanes of the type ArS_xH might be expected also from the reaction of ArSH with liquid elemental sulfur. The independent synthesis of various unsymmetrical sulfanes of the ArS_xH type (4) permitted the assignment of these new NMR signals and clearly indicated that not only HS_xH, but also unsymmetrical polysulfanes ArS_xH are generated *in situ* when an aromatic thiol reacts with molten sulfur. The behavior of the intensities of the signals assigned to HS_xH and ϕS_xH, and hence the concentrations of the species as a function of time are presented in detail in Figures 8 and 9.

Figure 8 compares the accumulation and disappearance of the intermediate products with the induction period observed in the second order kinetic analysis of the thiophenol–sulfur reaction. Inspection of Figure 8 shows that the HS_xH and ArS_xH species reach their maximum concentration at a point near the termination of the induction period.

Figure 9 shows the characteristic curves for the accumulation of these species in the reaction between thiophenol and liquid elemental sulfur at thiol–sulfur ratios of 1:18.8, 1:8, and 1:3 at 130°C. Several facts are qualitatively apparent from Figures 8 and 9. Irrespective of the relative amounts of HS_xH and ϕS_xH the concentration of HS_xH becomes dominant compared with the other intermediate species HS_xH when the relative concentration of ϕSH to sulfur is increased in the reaction mixture. The buildup to a maximum concentration of intermediates is associated with the induction period. Similar behavior was observed in the NMR spectra of reaction mixtures of other thiophenols with sulfur. However, such intermediates were not observed to form in the cases of *p*-amino and *p*-nitrothiophenol, further suggesting that these thiols react by a different mechanism.

End Products

The end products were obtained by carrying out runs under usual kinetic conditions. Vigorous evolution of H_2S occurred after the induction period. After the mercaptan was consumed completely, the residual mixture was extracted several times with ethanol and diethyl ether. In experiments with thiophenol, diphenyldisulfane and an oily residue were obtained as products. Diphenyldisulfane was synthesized independently, and the product proved identical to that produced by the reaction shown in Equation 6. The oily residue was subjected to NMR analysis and was found to be a mixture of diphenylpolysulfanes, ϕ–S_x–ϕ.

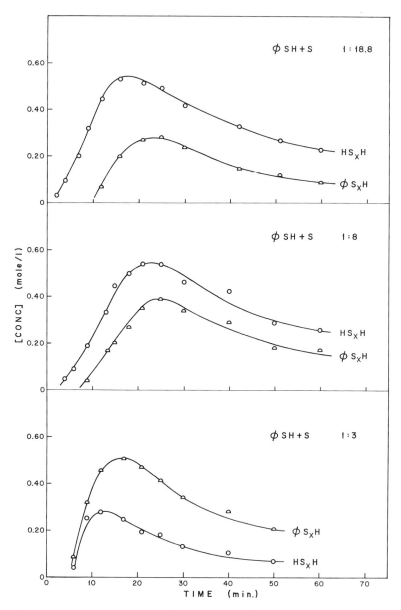

Figure 9.   Concentration of HS_xH and ArS_xH as a function of time for various thiophenol/sulfur ratios (130°C)

In experiments with *p*-methylthiophenol the presence of tri- and tetrasulfanes in the end products was confirmed by isolating these compounds. The structure of these sulfanes was substantiated by mixed melting point determinations with authentic samples independently synthesized (4). In experiments with other substituted thiophenols, NMR spectral analysis showed that the end products had the same general structure.

The Mechanism

The observation of an induction period, the inhibiting effect of radical scavengers, and the ease of rupture of cyclooctasulfur (S_8^R) to a catena-octasulfur ($\cdot S_8^{CH} \cdot$) biradical (7, 8) argue in favor of a radical initiated mechanism for the reaction of all but the *p*-amino and *p*-nitrothiophenols studied. The rate law described in Equation 5 is overall fifth order indicating that the mechanism is complex, involving several steps, some of which may be pre-rate determining equilibria. The second order dependence on thiol concentration is not surprising since the final product ArS_xAr requires the combination of two initial reactants. The third order dependence on sulfur, however, is accounted for less easily in mechanistic terms. Equations 7 and 8 represent an overall mechanism consistent with the facts considered above.

$$(n + 1)S_8^R \overset{K}{\rightleftharpoons} \cdot S_8^{CH} \cdot nS_8^R \tag{7}$$

$$\cdot S_8^{CH} \cdot nS_8^R + 2\,ArSH \xrightarrow{k_2} \underset{\substack{\text{sequence of} \\ \text{steps}}}{} \underset{\text{final products}}{ArS_xAr + H_2S} \tag{8}$$

Equations 7 and 8 yield an overall rate expression

$$-\frac{d[ArSH]}{dt} = k_2 K [RSH]^2 [S_8^R]^{n+1} \tag{9}$$

which is in accord with the experimental rate law described in Equation 5. Comparison with Equation 5 also establishes that $n = 2$ in the mechanism outlined in Equations 7 and 8. This implies that the biradical catena–octasulfur produced in the pre-rate determining equilibrium is "complexed" with two other cyclooctasulfur molecules. Other evidence for the existence of such a complex exists. According to Schenk and Thummler (9) pure liquid sulfur at approximately 130°C is 94.5% in the form of cyclooctasulfur, S_8^R, as shown by freezing point lowering measurements. The remaining 5.5% is thought to exist as catena–octasulfur,

·$S_8{}^{CH}$·. ESR measurements do not show the presence of any free radical species at this temperature. Attempting to rationalize the absence of ESR signals in liquid sulfur in this temperature range, Wiewiorowski and Touro (1) proposed that the odd electrons in ·$S_8{}^{CH}$· interact strongly through a system of d orbitals. They are delocalized over all sulfur atoms in the chain (10, 1). The electron density in the ·$S_8{}^{CH}$· species is used as a basis for introducing a theory of acid–base interaction in liquid sulfur. According to these authors, the chains having an excess electron density are considered to behave as Lewis bases, while the cyclooctasulfur species are thought to be Lewis acids. As shown above, this theory is used in the proposed mechanism and is rationalized in terms of Equation 7. The octatomic chains formed in the above system are solvated by the excess octatomic sulfur rings present. Using Schenk and Thummler's data, Wiewiorowski and Touro estimated that the ratio of Lewis acid ($S_8{}^R$) to Lewis base molecules (·$S_8{}^{CH}$·) in the complex was two, which is the same value found in the present kinetic investigation.

Because of the involvement of a pre-rate determining equilibrium and the possibility of more than one rate determining step being included in the subsequent sequence of steps, the enthalpy and entropy of activation determined in this study cannot be associated directly with a single molecular process. At the least the ΔH^* determined experimentally is a composite of the enthalpy change associated with the equilibrium shown in Equation 7 and one activation enthalpy. These comments apply equally to the entropy term. Despite such complications the value of 31.7 ± 1.2 kcal obtained for ΔH^* in this work is, within experimental error, in agreement with values of 32 and 32.8 kcal reported for the sulfur–sulfur bond cleavage in S_8 ring opening (7, 8). While positive values ranging up to +23 eu have been reported for the entropy of $S_8{}^R$ opening (7), the +7 eu value determined for ΔS^* in this study is in keeping with a dominant positive entropy term in the pre-rate determining ring opening.

We now consider the significance of the effect of varying substituents on the thiophenol. Recognizing that the p-amino and p-nitro substituted thiophenols do not fit a Hammett linear free energy relationship described by the other substituted thiols, an attempt is made to account for the experimentally observed value of −1.2 for the Hammett reaction constant ρ. A negative value for ρ is expressed in molecular terms as meaning that the reaction is favored by substituents which donate electrons to the reaction center. Such effects of substituents on radical reactions have been successfully discussed in terms of reaction center polarity. Russell (11) has suggested a polar transition state for the attack of peroxide radicals on phenols as shown in Equation 10 and by analogy, Equation 11 is suggested for the reaction under study here.

$$\overset{\delta_+ \quad \delta_-}{Z-C_6H_4-OH + \cdot O-O-R \rightarrow [Z-C_6H_4-O\cdot H:O-O-R] \rightarrow}$$
$$Z-C_6H_4O\cdot + HOOR \tag{10}$$

$$\overset{\delta_+ \quad \delta_-}{Z-C_6H_4-SH + \cdot S-S-R \rightarrow [Z-C_6H_4-S\cdot H:S-S-R] \rightarrow}$$
$$Z-C_6H_4S\cdot + HSSR \tag{11}$$

In the above reaction the S–H bond becomes polarized giving a partial positive charge on the sulfur (δ_+) with an electron being transferred partially to the attacking radical (δ_-). Substituents in the thiol molecule which favor the polar structures preferred by the attacking radical will speed the reaction. Values of ρ ranging from -0.4 to -1.5 have been observed for radical abstraction of hydrogen in reactions of this kind (*11, 12, 13, 14*). The $\rho = -1.2$ value observed here suggests that there is a moderate degree of charge separation in the rate determining transition state involving hydrogen abstraction from the thiol by an $ArS_x\cdot$ or $HS_x\cdot$ radical.

It is appropriate now to consider the different behavior of both the *p*-amino and *p*-nitrothiophenols. The accumulation of evidence suggests that these two thiols react by a different mechanism. It is known that amines have a marked effect on the opening of $S_8{}^R$ and react with sulfur rapidly to form ions (*15, 16*). Davis (*17*) has proposed a scheme for the initial steps in the reaction of sulfur and amines. Probably, the *p*-aminothiophenol reacts with sulfur by an ionic rather than a free radical mechanism. The question remains, however, as to why *p*-nitrothiophenol reacts in a similar manner to the *p*-amino compound. The reduction of nitro groups to amino functions by hydrogen sulfide and polysulfides is well known (*18*); hydrogen sulfide is a reaction product in the thiol/liquid sulfur reaction. Examination of the NMR spectra recorded during the reaction of *p*-nitrothiophenol and sulfur shows that small amounts of products having an amino group are formed during the reaction. Traces of such amines would suffice to catalyze the ionic reaction mechanism for *p*-nitrothiophenol.

The concentration–time behavior of the HS_xH and ArS_xH intermediates presented in Figures 8 and 9 permits further comments on the details of the sequence of steps following the radical initiation. It is not possible to elaborate fully the sequence of steps summarized (Equation 8), but a few limiting criteria can be established. The fact that the concentrations of the HS_xH and ArS_xH intermediates maximize during the induction period suggests that they are the immediate and relatively stable initial products of reaction of the propagating radical species. It is suggested that the initial complexed catenasulfur biradical (henceforth $\cdot S_x \cdot$) reacts initially with thiol to yield a sequence of steps shown in Equations 12 to 15.

$$ArSH + \cdot S_x \cdot \rightleftharpoons ArS \cdot + HS_x \cdot \tag{12}$$

$$HS_x \cdot + ArSH \rightleftharpoons ArS \cdot + \underline{HS_xH} \tag{13}$$

$$ArS \cdot + nS_8 \rightleftharpoons ArS_x \cdot \tag{14}$$

$$ArS_x \cdot + ArSH \rightleftharpoons ArS \cdot + \underline{ArS_xH} \tag{15}$$

The hydrogen polysulfides or sulfanes have been shown (3) to be unstable under the reaction conditions used and to undergo decomposition by disproportionating via other sulfanes to hydrogen sulfide and sulfur (Equation 6). The aryl sulfanes (ArS_xH) are generally less stable than the HS_xH species and are expected to decompose in a similar manner yielding the diaryldisulfane and hydrogen sulfide (Equation 17).

$$HS_xH \rightarrow H_2S + nS_8 \tag{16}$$

$$ArS_xH \rightarrow Ar_2S_2 + nS_8 \tag{17}$$

Conclusion

The reaction of aromatic thiols with liquid sulfur in the temperature range between the melting point of sulfur and the equilibrium polymerization temperature is free radical in nature and similar to the reaction of hydrogen sulfide with sulfur under the same conditions. Intermediates of the HS_xH and ArS_xH type are formed in the reaction but decompose to form mainly H_2S, Ar_2S_2, and sulfur. Kinetic evidence suggests that the initial sulfur radical species formed is complexed with two cyclooctasulfur molecules. While this idea agrees with previous suggestions based on non-kinetic evidence it is not conclusive. p-Amino and p-nitrothiophenol react by a mechanism that is probably ionic initiated by the catalytic effect of the amino group present or generated in situ by H_2S reduction of the nitro group.

Literature Cited

(1) Wiewiorowski, T. K., Touro, F. J., J. Phys. Chem. (1966) 70, 3528.
(2) Hyne, J. B., Muller, E., Wiewiorowski, T. K., J. Phys. Chem. (1966) 70, 3733.
(3) Muller, E., Hyne, J. B., J. Amer. Chem. Soc. (1969) 91, 1907.
(4) Langer, H. J., Hyne, J. B., Can. J. Chem., to be published.

(5) Jaffé, H. H., *Chem. Rev.* (1953) **53**, 191.
(6) Wiewiorowski, T. K., Touro, F. J., *J. Phys. Chem.* (1966) **70**, 234.
(7) Tobolsky, A. V., Eisenberg, A., *J. Amer. Chem. Soc.* (1959) **81**, 780.
(8) Wiewiorowski, T. K., Parthasarathy, A., Slaten, B. L., *J. Phys. Chem.* (1968) **72**, 1890.
(9) Schenk, P. W., Thümmler, U., *Z. Electrochem.* (1959) **63**, 1002.
(10) Miller, J. D., "A Theoretical Investigation of the Nature of the Bonding in Sulfur Containing Molecules," Thesis, Tulane University, 1970.
(11) Russell, G. A., *J. Amer. Chem. Soc.* (1956) **78**, 1047.
(12) Ingold, K. U., *J. Phys. Chem.* (1960) **64**, 1636.
(13) Schaafsma, Y., Bickel, A. F., Kooyman, E. C., *Rec. Trav. Chim.* (1957) **76**, 180.
(14) Walling, C., Miller, B., *J. Amer. Chem. Soc.* (1957) **79**, 4181.
(15) Jennen, A., Hens, M., *Comp. Rend.* (1956) **242**, 786.
(16) Mayer, R., Gewald, K., *Angew. Chem.* (1967) **79**, 298.
(17) Davis, R. E., Nakshbendi, H. F., *J. Amer. Chem. Soc.* (1962) **84**, 2085.
(18) Schröter, R., in Houben-Weyl-Müller: "Methoden der Organischen Chemie," 4th ed., Vol. XI/1, Thieme-Verlag, Stuttgart, 1957.

RECEIVED March 5, 1971.

9

Reactions of the Hydrogen Atom

VI. The Photolysis of Thiols.

J. P. STANLEY, R. W. HENDERSON, and WILLIAM A. PRYOR

Louisiana State University, Baton Rouge, La. 70803

We have generated hydrogen atoms by photolysis of either thiols or tert-butyl peroxyformate, and for each compound we have devised two different kinetic schemes to study the reaction in which these H atoms abstract hydrogen from organic compounds in solution. The data from all four kinetic systems are in good agreement. The results also agree with data from aqueous radiolysis for almost all compounds studied. The results are discussed, and the selectivity of the hydrogen atom is compared with that of alkyl and aryl radicals.

Photolysis of thiols and disulfides produces thiyl radicals in a reaction with a high quantum yield (1).

$$RSH \xrightarrow{h\nu} RS\cdot + H\cdot$$

$$RSSR \xrightarrow{h\nu} 2RS\cdot$$

For thiols, the expected other product, H atoms, is produced, and in the gas phase they are "hot" as initially formed ($2, 3, 4$). Simple thiols give H_2 with a quantum yield of near unity ($2, 5$). The photolysis of sulfides and disulfides in the liquid phase has been studied also, and thiyl radicals are primary products in these reactions as well (6). A detailed study of the photochemistry of liquid thiols is underway in these laboratories (7), and preliminary results indicate that thiyl radicals and hydrogen atoms are the main products of photolysis.

The H atom is an important species in radiation biology. Radiolysis of aqueous, organic, or biological systems results in the formation of various free radicals, including the hydrogen atom. Although several techniques have been developed to separate the reactions of the hydrogen atom from the reactions of the other reactive species produced by radiation (8, 9), it is desirable to generate H atoms by a non-radiolytic technique. Some years ago (10, 11) we began a detailed study of the photolysis of thiols intending to use this technique to study the reactions of the H atom in solution.

In contrast with the vapor phase, in the liquid phase the hydrogen atom is thermalized before reaction. Typical hydrogen abstraction reactions of hydrogen atoms have rate constants of the order of 10^5–10^8 M^{-1} sec^{-1} (12), considerably below the diffusion controlled limit. Therefore hydrogen atoms produced by photolysis undergo sufficient collisions to become thermalized prior to reaction. At least one study of thiol photolysis in the gas phase concludes that the reactions of the hydrogen atoms produced there are independent of exciting wavelength (2).

As our research progressed, it became clear that we should aim at developing more than one method for the generation and kinetic study of H atoms so that our results could be intercompared and their reliability assessed. We now have developed four kinetic systems; two involve the photolysis of thiols, and two involve the photolysis of *tert*-butyl peroxyformate, $HCO_3C_4H_9$–*tert* (BUP). The data for all four systems are in agreement, and this fact, as well as internal controls and tests, suggests that H atoms are produced in all four systems as discrete kinetic intermediates. We have compared also our results with radiolysis data; generally, the agreement is excellent.

Data

Method A. The first successful system we developed for studying the reactions of the hydrogen atom involves the photolysis of deuterated thiols to produce deuterium atoms (11, 13, 14). The deuterium atoms then abstract either deuterium or hydrogen from the thiols or hydrogen from an organic hydrogen donor, QH. The reaction scheme is shown below.

Method A:

$$RSD \xrightarrow{h\nu} RS\cdot + D\cdot \qquad (1)$$

$$D\cdot + RSD \rightarrow RS\cdot + D_2 \qquad (2)$$

$$D\cdot + RSD \rightarrow \cdot RSD + HD \qquad (3)$$

$$D\cdot + QH \rightarrow Q\cdot + HD \qquad (4)$$

$$Q\cdot + RSD \rightarrow QD + RS\cdot \qquad (5)$$

$$2RS\cdot \rightarrow RSSR \qquad (6)$$

where ·RSD is a thiol that has lost a hydrogen atom from its alkyl group, and the other symbols have their usual meaning. At low conversions, Reactions 5 and 6 can be neglected in the kinetic analysis since these reactions do not affect the yields of HD or D_2.

Kinetic anlysis of this system indicates that a plot of $[HD]/[D_2]$ vs. $[QH]/[RSD]$ should yield a straight line of slope $k_H I_4/k_2$, where $I_4 = k_D/k_H$ for hydrogen or deuterium atoms reacting with QH. Values of I_4 are nearly unity and are essentially the same for all QH compounds (15, 16, 17, 18, 19).

We have measured the relative reactivities of 15 organic hydrogen donors by this method using thiophenol-d and 2-methyl-2-propanethiol-d. The two different thiols agree within about 10%. Since the bond strengths of the S–H bond in the two thiols differ by about 13 kcal (20, 21), this agreement indicates that no reaction which involves breaking or forming a bond to sulfur is interfering with our analysis.

Method B. Another, similar system we have used to measure the relative reactivities of hydrogen atoms involves the use of tritiated thiol. In this case, the rate of incorporation of tritium into the QH is measured (10, 22). This system is shown below:

Method B:

$$RSH(T) \xrightarrow{h\nu} RS\cdot + H\cdot(T\cdot) \tag{7}$$
$$H\cdot + QH \rightarrow Q\cdot + H_2 \tag{8}$$
$$H\cdot + Q'H \rightarrow Q'\cdot + H_2 \tag{9}$$

$$Q\cdot + RSH(T) \rightarrow QH(T) + RS\cdot \tag{10}$$

$$Q'\cdot + RSH(T) \rightarrow Q'H(T) + RS\cdot \tag{11}$$

$$Q\cdot + Q'H \rightarrow QH + Q'\cdot \tag{12}$$

The relative specific activities of recovered $QH(T)$ and $Q'H(T)$ are related directly to the relative rates of Equations 8 and 9. One hydrogen donor is selected as the standard, and the reactivities of other compounds are determined by comparison of their specific activities with that of the standard. There is an isotope effect on the steps in which activity is incorporated in the hydrogen donor (Equations 10 and 11). The magnitude of these isotope effects varies with the nature of the organic radical $Q\cdot$, but we have now measured the value of the isotope effect on Equation 10 for a large number of such $Q\cdot$ radicals, and the raw data could be corrected for this small variation (23). The reaction given in Equation 12, which potentially could lead to erroneous results, is unimportant when excess thiol is present.

Table I.   Relative Rate Constants for Reaction of Hydrogen Atoms
with Hydrogen Donors

	Method[e]				Radiolysis	
QH	A[a]	B[b]	C[c]	D[d]	Foot-note f	Foot-note g
Hexane	(1)	(1)	(1)	(1)		(1)
Nonane	2.2	1.4				1.7
Dodecane	2.3	2.0				
2,3-Dimethylbutane	2.2	2.1	2.1	2.4		3.2
2,4-Dimethylpentane	1.5	0.84				3.6
2,5-Dimethylhexane	2.4	1.2				
Cyclopentane	1.2	0.87				1.1
Cyclohexane	1.2	1.1	1.4			
Methanol	0.42	0.43	0.34	.39	(0.42)	
Ethanol	1.2	0.86	1.4	1.5	4.2	
2-Propanol	1.9	1.8	2.6	2.8	13.0	
2-Methyl-2-propanol	0.03	<0.1	0.05	0.03	0.03	
Dioxane	3.0	1.2	2.2		1.4	
Tetrahydrofuran	9.1		8.2		7.7	
Isopropyl ether	4.8	2.1	4.9			

[a] Relative values of $k_H I_4$ obtained from the deuterated thiol system; see text.
[b] Relative values of $k_H I_{10}$ obtained from the tritiated thiol system; the isotope effect in this system tends to compress the scale of reactivities slightly, see text.
[c] Relative values of k_H obtained from the *tert*-butyl performate–thiol system; see text.
[d] Relative values of k_H obtained from the *tert*-butyl performate–benzene system; see text.
[e] Most of these values are believed to be accurate to at least ±20%.
[f] Average values of data reviewed by M. Anbar and P. Neta (*12*).
[g] Data of T. J. Hardwick (*26*).

Method C. The third system we have used to measure hydrogen atom reactivities involves the photolysis of *tert*-butyl peroxyformate (BUP), $H–CO_2–O–C(CH_3)_3$, to produce hydrogen atoms (*24*). The kinetic system involves a competition in which the H atoms either abstract deuterium from thiol-*d* (added as a standard reactant) or from QH. The reactions are shown below.

Method C:

$$BUP \xrightarrow{h\nu} H\cdot + \text{other products} \tag{13}$$

$$H\cdot + QH \xrightarrow{k_H} H_2 + Q\cdot \tag{14}$$

$$H\cdot + RSD \rightarrow H_2 + \cdot RSD \tag{15}$$

$$H\cdot + RSD \rightarrow HD + RS\cdot \tag{16}$$

This system is quite similar kinetically to Method A, but it is superior because it involves the hydrogen atom instead of the deuterium atom.

Method D. Our fourth system uses *tert*-butyl performate in a thiol free system to generate hydrogen atoms. Benzene is used as a hydrogen atom scavenger, and the absolute yield of hydrogen is measured when varying amounts of the hydrogen donor QH are added.

Method D:

$$H\cdot + QH \rightarrow Q\cdot + H_2 \tag{17}$$

$$H\cdot + PhH \rightarrow (PhH_2)\cdot \tag{18}$$

The results of this system agree with our other systems (*25*). This indicates that the other systems, which all contain thiol either as the H atom precursor or as the standard reactant, do not have a systematic error as a result of, for example, the presence of the thiyl radical. The agreement of Method D with the other systems also eliminates the possibility that it is a photoexcited thiol which is the immediate precursor of H_2, rather than H atoms themselves.

Discussion

Table I lists the data obtained by our four methods for the reactivities of various organic compounds (relative to hexane) and also some of the previous data collected by radiolysis techniques. All of the data agree well with the exception of the reactivities of the alcohols as measured by aqueous radiolysis and the more reactive hydrocarbons as measured by radiolysis. The disagreement between our work and radiolysis for ethanol and 2-propanol cannot be attributed to experimental error since numerous laboratories have studied the radiolysis of these alcohols. The reason for this discrepancy is unknown at present. However, we feel that two arguments suggest that our data may be correct and that the radiolysis results may contain some subtle systematic error for ethanol and 2-propanol. The first is discussed below in connection with Table II. The second is the lack of internal consistency of the radiolysis data themselves. When compared with other compounds studied by radiolysis which contain RCH_2OH or R_2CHOH groups, ethanol and 2-propanol stand out as unusually reactive. This point is discussed by Pryor and Stanley (Table IV of Reference *4*).

Since the submission of this paper, the measurement of a large number of values of k_H for H-atoms by a new radiolysis technique has been reported. Neta, Fessenden, and Schuler have studied the alcohols, and they conclude that the published values for ethanol and isopropyl alcohol are correct. However, they point out that these values are not consistent with values for other OH-containing substrates (*27*).

Table II lists the relative rate constants for hydrogen abstraction from alkanes and alcohols by alkyl and aryl radicals. As can be seen, the relative rates of abstraction of primary, secondary, and tertiary hydrogens from alkanes by the various radicals are about the same, with the exception of the radiolysis values for the H atom. A similar observation is made

Table II.   Rate Constants for Hydrogen Abstraction from Organic Compounds by Various Radicals

Relative Reactivities Per Hydrogen

Hydrogen Donor	$H \cdot$ Best Data of Pryor et al.[a] soln 35°	Rad.[b] soln 25°	Rad.[c] gas 25°	$Me \cdot$ gas 182°	$Me \cdot$ soln 110°,25°	$RCH_2 \cdot$ [d] soln 130°	$Ph \cdot$ [e] soln 60°
Alkanes							
Primary	(1)		(1)	(1)[f]	(1)[g]	(1)	1
Secondary	3.3		7	5[f]	4.3[g]	3	9
Tertiary	31		88	33[f]	46[g]		44
Alcohols							
CH_3OH	(1)	(1)		(1)[h]	(1)[i]	(1)	(1)
CH_3CH_2OH	4	15		5[h]	5[i]	5	3
$(CH_3)_2CHOH$	17	92		25[h]	46[i]	20	9

[a] Data taken from Methods A, B, and C; see text.
[b] Data taken from a review by M. Anbar and P. Neta (12).
[c] Data of T. J. Hardwick (26).
[d] G. A. Mortimer (28).
[e] R. F. Bridger and G. A. Russell (29).
[f] W. M. Jackson, J. R. McNesby, and B. deB. Darwent (30).
[g] W. A. Pryor, D. Fuller, and J. P. Stanley (31).
[h] A. A. Herrod; data are for the $CD_3 \cdot$ radical (32).
[i] J. K. Thomas (33).

for reactions with alcohols. The similarity in selectivity profiles for the H atom and alkyl radicals is partially a result of the approximately equal bond strengths of the bonds being formed. The reason for the similarity of the reactivity pattern for phenyl radicals is not entirely clear.

The data of Table II support the suggestion that there may be a systematic error in the values for 2-propanol and ethanol as determined by aqueous radiolysis. Our reactivity profile for the H atom is closer to what would be expected in comparison with other radicals than is the radiolysis data profile.

Acknowledgments

The generous support of the National Institutes of Health, U. S. Public Health Service (Grant No. GM-11908) is gratefully acknowledged.

William A. Pryor wishes to express his gratitude to W. F. Libby and the UCLA Chemistry Department for hospitality during his tenure as John Simon Guggenheim Fellow at UCLA, September 1970–April 1971, and to Professor Melvin Calvin for hospitality while at Berkeley, April–August 1971.

Literature Cited

(1) Pryor, W. A., Stanley, J. P., Lin, T. H., *Quart. Rep. Sulfur Chem.* (1970) **5**, 305, and references cited therein.

(2) Steer, R. P., Knight, A. R., *Can. J. Chem.* (1968) **46**, 2878.

(3) Sturm, G. P., Jr., White, J. M., *J. Chem. Phys.* (1969) **50**, 5035.

(4) White, J. M., Sturm, G. P., Jr., *Can. J. Chem.* (1969) **47**, 357.

(5) Steer, R. P., Knight, A. R., *J. Phys. Chem.* (1968) **72**, 2145.

(6) Sayomol, K., Knight, A. R., *Can. J. Chem.* (1968) **46**, 999.

(7) Davis, W. H., unpublished results.

(8) Ausloos, P., Ed., "Fundamental Processes in Radiation Chemistry," Interscience Publishers, New York, 1968.

(9) Stein, G., "Radiation Chemistry of Aqueous Systems," Weizmann Science Press of Israel, Jerusalem, and Interscience Publishers, New York, 1968.

(10) Pryor, W. A., Griffith, M. G., *J. Amer. Chem. Soc.* (1971) **93**, 1408.

(11) Pryor, W. A., Stanley, J. P., Griffith, M., *Science* (1970) **169**, 181.

(12) Anbar, M., Neta, P., *Int. J. Appl. Radiat. Isotopes* (1967) **18**, 493.

(13) Pryor, W. A. Stanley, J. P., *Intra-Sci. Chem. Rep.* (1970) **4**, 99.

(14) Pryor, W. A., Stanley, J. P., *J. Amer. Chem. Soc.* (1971) **93**, 1412.

(15) Anbar, M., Meyerstein, D., in "Radiation Chemistry of Aqueous Systems," p. 116, G. Stein, Ed., Interscience, New York, 1968.

(16) Boato, G., Careri, G., Cimino, A., Molinari, E., Valpi, G. C., *J. Chem. Phys.* (1956) **24**, 783.

(17) Hirschfelder, J., Eyring, H., Topler, B., *J. Chem. Phys.* (1936) **4**, 170.

(18) Schulz, W. R., LeRoy, D. J., *J. Chem. Phys.* (1965) **42**, 3869.

(19) Westenberg, A. A., deHaas, N., *J. Chem. Phys.* (1967) **47**, 1393.

(20) Kerr, J. A., *Chem. Rev.* (1966) **66**, 465.

(21) Mackle, H., *Tetrahedron* (1963) **19**, 1159.

(22) Pryor, W. A., Lin, T. H., unpublished.

(23) Pryor, W. A., Kneipp, K., *J. Amer. Chem. Soc.* (1971) **93**, 5584.

(24) Pryor, W. A., Henderson, R. W., *J. Amer. Chem. Soc.* (1970) **92**, 7234.

(25) Pryor, W. A., Henderson, R. W., submitted for publication.

(26) Hardwick, T. J., *J. Phys. Chem.* (1961) **65**, 101.

(27) Neta, P., Fessenden, R. W., Schuler, R. H., *J. Phys. Chem.* (1971) **75**, 1645.

(28) Mortimer, G. A., *J. Polymer Sci., Part A-1* (1966) **4**, 88.

(29) Bridger, R. F., Russell, G. A., *J. Amer. Chem. Soc.* (1963) **85**, 3754.

(30) Jackson, W. M., McNesby, J. R., Darwent, P. deB., *J. Chem. Phys.* (1962) **37**, 1610.

(31) Pryor, W. A., Fuller, D., Stanley, J. P., *J. Amer. Chem. Soc.*, in press.

(32) Herrod, A. A., *Chem. Commun.* (1968) 891.

(33) Thomas, J. K., *J. Phys. Chem.* (1967) **71**, 1919.

RECEIVED March 5, 1971.

The Addition of Sulfur Atoms to Olefins

O. P. STRAUSZ

University of Alberta, Edmonton, Alberta, Canada

Absolute rate coefficients and Arrhenius parameters have been obtained for the cycloaddition reaction of $S(^3P_{2,1,0})$ atoms with a representative series of olefins and acetylenes. The activation energies are small, and they exhibit a trend with molecular structure which is expected for an electrophilic reagent. The A-factors show a definite trend which can be attributed to steric repulsions and a generalized secondary α-isotope effect explained by activated complex theory. Secondary α-H/D kinetic isotope effects have been measured and their origin discussed. Hartree-Fock type MO calculations indicate that the primary product of the $S(^3P)$ + olefin reaction is a ring-distorted, triplet state thiirane, with a considerable energy barrier with respect to rotation around the C–C bond.

Sulfur atoms and the atoms of the group VIa elements undergo facile addition to olefins and acetylenes. These reactions represent the simplest examples of a cycloaddition reaction. Among them, the sulfur atom reactions stand out with their experimental simplicity, cleanliness, and unique reactivity. Therefore the sulfur atom–olefin system is a model for studying some of the little explored aspects of cycloaddition reactions. Since the first report on sulfur atom addition reactions which appeared in 1962, a considerable amount of quantitative data and relevant information have accumulated on the subject, which are summarized briefly here.

In the ground electron configuration of the group VIA elements four electrons are distributed over three p-orbitals. This distribution gives rise to five spectroscopic states designated as $^3P_{2,1,0}$, $^1D_2$, and $^1S_0$. The lowest lying state is the $^3P_2$ state, though the spin-orbital splitting is small, only 1.6 kcal for sulfur. Radiative transitions among these states are forbidden by rigid selection rules; therefore the excited atoms have long lifetimes and are able to undergo bimolecular reactions.

To date, studies have been confined largely to the reactions of ground state triplet and the lowest excited $^1D_2$ states of these atoms. Possible differences in reactivities among the triplet components have been ignored.

For sulfur atoms in all kinetic studies reported, the source was the *in situ* photolysis of carbonyl sulfide or carbon disulfide, which proceed according to the equations:

$$COS + h\nu \underset{\lambda \leqslant 2800\text{A}}{\rightarrow} CO + S(^1D_2) \tag{1}$$

$$CS_2 + h\nu \underset{\lambda \leqslant 2100\text{A}}{\rightarrow} CS + S(^3P_{2,1,0}) \tag{2}$$

Thus, Reaction 1 produces sulfur atoms in their lowest excited singlet-D_2 state with 26.4 kcal excitation energy while Reaction 2 results in their ground, triplet-$P_{2,1,0}$ states. Alternative sources of triplet sulfur atoms are the triplet mercury photosensitization of COS and the direct photolysis of COS in the presence of excess CO_2. Carbon dioxide is a quencher of singlet, excited sulfur atoms

$$S(^1D_2) + CO_2 \rightarrow S(^3P_J) + CO_2, \tag{3}$$

with a rate constant, $k_3 > 1 \times 10^{10}$ liters mole^{-1} sec^{-1} (*1*).

Both $S(^1D_2)$ and $S(^3P)$ atoms react readily with olefins and acetylenes, but they exhibit different reactivities.

$S(^1D_2)$ atoms afford two principal products with olefins, thiirane, and mercaptan. Thiirane comes from the cycloaddition of the sulfur atom across the olefinic double bond. Mercaptans, which contain vinylic- and alkenyl-types, are characteristic insertion products formed upon a concerted, single step attack of the $S(^1D_2)$ atom on the CH bond. The insertive ability of the $S(^1D_2)$ atom has been demonstrated in separate studies. Vinylic type mercaptans are produced only from terminal olefins, and their formation may be related also to the isomerization of the chemically activated episulfides, $CH_2\underset{\diagdown \quad \diagup}{\overset{}{\underset{S}{}}}CH_2{}^* \rightarrow CH_2 = CHSH$, con-

taining an excess energy of $\sim$85 kcal. The feasibility of this step is based on thermal studies of episulfides at elevated temperatures.

The thiirane-forming cycloaddition follows a stereospecific path. This reaction has been shown for three pairs of olefins (*2, 3*), *cis-trans*-butene-2, *cis-trans*-1,2-dichloro- and difluoroethylenes.

The reactivity of olefins with respect to $S(^1D_2)$ atom increases as the number of alkyl substituents on the doubly bonded carbon atoms increase. The approximate values of rate constants for the reactions

$$S(^1D_2) + C_2H_4 \rightarrow CH_2\underset{S}{\diagdown\diagup}CH_2 \qquad (4a)$$

$$\rightarrow S(^3P) + C_2H_4 \qquad (4b)$$

$$S(^1D_2) + C_3H_6 \rightarrow CH_3CH\underset{S}{\diagdown\diagup}CH_2 \qquad (5a)$$

$$\rightarrow S(^3P) + C_3H_6 \qquad (5b)$$

$$S(^1D_2) + i\text{-}C_4H_8 \rightarrow (CH_3)_2C\underset{S}{\diagdown\diagup}CH_2 \qquad (6a)$$

$$\rightarrow S(^3P) + i\text{-}C_4H_8 \qquad (6b)$$

are (*1*, *4*):

$$k_4 \sim 2 \times 10^{10} \text{ liters mole}^{-1} \text{ sec}^{-1}$$

$$k_5 \sim 6 \times 10^{10}$$

$$k_6 \sim 15 \times 10^{10}$$

The activation energy of the addition reactions is small, 1 kcal or less. The mercaptan-forming steps have about 0.5 kcal higher activation energies.

Ground state, triplet sulfur atoms give only thiirane with olefins, and the yields with the simple, less reactive olefins are nearly ideal. The unique feature of the reaction is that it follows a stereospecific path and thereby provides the first example of a stereospecific cycloaddition of a divalent triplet state reagent.

Quantitative studies of $S(^3P)$ atom reactions have been carried out with about two dozen olefins (*5*, *6*). The rate coefficients and Arrhenius parameters are summarized in Table I. The absolute rate coefficients were determined in flash photolysis experiments using kinetic absorption spectrometry (*6*, *7*). Mixtures of 0.1 torr COS and 200 torr CO_2 were flash photolyzed in the presence of an olefin, and $S(^3P)$ atoms concentrations were monitored by measuring the optical densities of the 1807 ($^3P_2$) and 1820A ($^3P_1$) atomic transitions.

The trend in activation energies (Table I) shows the electrophilic nature of attack by the sulfur atom; increasing number of alkyl substituents on the doubly bonded carbons decreases the value of E_a, while increasing number of halogen substituents on the doubly bonded carbons increases the value of E_a. These variations are correlated with molecular properties such as ionization potentials, excitation energies, and bond

Table I.   Arrhenius Parameters for the Addition of $S(^3P_J)$ Atoms to Olefins

	$E_a(C_2H_4) - E_a{}^a$ (kcal/mole)	$A/A(C_2H_4)^a$	k^b (liter mole^{-1} sec^{-1}) $\times 10^9$
C_2H_4	0.0	1.0	0.9 ± 0.1
C_2D_4	0.0	1.14	
CD_2CH_2	0.0	1.07	
cis-CHDCHD	0.0	1.04	
[structure]	1.14	1.0	6 ± 1
[structure]	1.72	0.75	9 ± 1
[structure]	2.09	0.53	
[structure]	2.01	0.65	12 ± 2
[structure]	2.36	0.97	45 ± 6
[structure]	3.01	0.51	
[structure]	3.36	0.50	77 ± 10
[structure]	2.04	2.4	
[structure]	2.83	0.78	
[structure]	2.15	0.67	
[structure]	−0.73	1.4	
[structure] F	−2.71	1.7	
[structure] F	−2.62	3.4	
C_2F_4	−1.4	1.4	
[structure] F	−1.5	1.2	
[structure] Cl	−0.52	3.4	
$S + S\triangleleft \rightarrow S_2 + C_2H_4$	1.8	8.3	
$S + S\triangleleft \rightarrow S_2 + C_3H_6$	2.1	8.4	

^{a} Temperature range extended from 25° to the decomposition point of the thiirane product ∼100–160°C.

^{b} T = 25°C.

orders in a similar fashion as for $O(^3P)$ atoms (8), the CF_3 radical (9, 10), and other electrophilic reagents (8). The activation energy for the ethylene reaction is 3.36 kcal larger than for the tetramethylethylene (TME) reaction. If the TME reaction is zero or positive, then $E_a(C_2H_4)$ > 3.4 kcal. The A-factor for the abstraction reaction—*i.e.*, $S + SCH_2CH_2$ → $S_2 + C_2H_4$—is 8.3-fold greater than that for the ethylene reaction and since the former cannot be higher than a few tenths of the gas kinetic collision frequency, the upper limit for $A(C_2H_4)$ is $\leqslant 10^{10}$ liters mole^{-1} sec^{-1}. (Since the corresponding values reported for $O(^3P)$ (11) and $Se(^3P)$ atom (12) reactions are $\sim 1 \times 10^{10}$ and 1.1×10^{10} liters mole^{-1} sec^{-1}, respectively, it is assumed that $A(C_2H_4) = 1 \times 10^{10}$ liters mole^{-1} sec^{-1}.) Therefore the upper limit for $k(C_2H_4) = 10^{10} e^{-3.360/RT}$. At room temperature this leads to a value of 3.7×10^7 liters mole^{-1} sec^{-1} while the absolute value from the flash experiments is 9×10^8 liters mole^{-1} sec^{-1}. The discrepancy is resolved by assuming that the activation energy for the TME reaction has a negative value of ~ 1.9 kcal. Recent measurements by Braun *et al.* (13), using flash photolysis with fluorescence spectroscopic techniques, yielded a value of $E_a(C_2H_4) = 1.5$ kcal in agreement with the present estimate. From existing data on the $O(^3P)$ (8, 11) and $Te(^3P)$ (6) atom reactions it also appears that the addition reaction of all the group VIa atoms to TME has a negative activation energy. This result is surprising since no precedent is known for a negative activation energy existing for an addition reaction in the second order region. The negative activation energy indicates that the addition is a nonadiabatic process. The atom and olefin initially approach on a potential energy surface with a shallow minimum. The repulsive part of this surface intersects an attractive potential surface, and the rate of the reaction is determined by the following factors:

(a) The potential energy difference between the intersection of the two surfaces and the separated reactants.

(b) The number of internal degrees of freedom in the reaction complex.

(c) The probability of crossing between the two surfaces.

Thus, a positive experimental activation energy results for the intersection of the two potential energy surfaces lying above the energy of the separated reactants since an increase in temperature enhances the rate of crossing, as with ethylene. If the intersection of the two potential surfaces lies below the energy of the separated reactants, an increase in temperature favors redissociation of the reaction complex to reactants, exhibiting a negative experimental activation energy for the product-forming reaction, as found in the reactions with TME.

A trend also appears in the variation of the A-factors. The involvement of two opposing effects is recognized. For alkyl substituents steric repulsion predominates, and the A-factor decreases slowly with increasing number of alkyl substituents. The lowest A-factor obtained is for TME, which has a value of one-half that for ethylene. For halogen atoms the steric repulsion is overcompensated by an enhancing effect, which is larger for chlorine than for fluorine. In the fluoroolefin series as the number of fluorine atoms increases steric repulsions take over, and the A-factor starts to decrease. The origin of this enhancing effect is discussed later.

Secondary α-H/D kinetic isotope effects were measured for three deuterated ethylenes. The values (Table I) are small, and they are inverted as expected.

From transition state theory, the following equation is derived for the kinetic isotope effect (14, 15) (Equation 1):

$$\frac{k_H}{k_D} = \left(\frac{\left(\frac{I_{A_2} I_{B_2} I_{C_2}}{I_{A_1} I_{B_1} I_{C_1}} \right)^{1/2} \left(\frac{M_2}{M_1} \right)^{3/2}}{\left(\frac{I^{\neq}_{A_2} I^{\neq}_{B_2} I^{\neq}_{C_2}}{I^{\neq}_{A_1} I^{\neq}_{B_1} I^{\neq}_{C_1}} \right)^{1/2} \left(\frac{M^{\neq}_2}{M^{\neq}_1} \right)^{3/2}} \right) \times \left(\frac{\prod\limits_{i}^{3N-6} \left[\frac{1 - \exp(-U_{1i})}{1 - \exp(-U_{2i})} \right]}{\prod\limits_{i}^{3N^{\neq}-7} \left[\frac{1 - \exp(-U^{\neq}_{1i})}{1 - \exp(-U^{\neq}_{2i})} \right]} \right)$$

$$\times \left(\frac{\exp\left[\sum\limits_{i}^{3N-6} (U_{1i} - U_{2i})/2 \right]}{\exp\left[\sum\limits_{i}^{3N^{\neq}-7} (U^{\neq}_{1i} - U^{\neq}_{2i})/2 \right]} \right) \tag{1}$$

where the $\neq$ signifies the transition state, 1 and 2 refer to the light and heavy molecules respectively, I's are the principal moments of inertia, M's the molecular masses, $u_i = hc\omega_i/kT$, ω_i is a normal vibrational frequency, and i is any integer.

The first factor of this equation is derived from the translational and rotational partition functions and is designated MMI. The second factor is the contribution of vibrational excitation and is denoted EXC, and the last factor contains the vibrational zeropoint energies and is designated ZPE. Thus, the equation is rewritten in the form

$$k_H/k_D = \text{MMI} \times \text{EXC} \times \text{ZPE}$$

The isotope effect is caused by the difference in the geometry and force constants between reactants and activated complexes at the positions of isotopic substitution. The magnitude of the isotope effect depends on the magnitude of these changes during passage from reactants to the activated complex, and an inverse secondary isotope effect, $(k_D/k_H > 1)$ requires an increase in the force constants.

The kinetic isotope effect (Equation 1) depends on the properties of the transition state. Since there is no way to determine these properties experimentally, a "reasonable" model is assumed. Calculations were performed on two different models for the structure of the activated complex. One of these consisted of the suprafacial, least-motion approach of the $S(^3P)$ atom to yield a symmetrical thiirane molecule, while the other was an asymmetrical approach at $\sim100°$ to the C–C bond to give a ring-distorted diradical-like intermediate (*vide infra*). Unfortunately, the calculation is not sensitive to the models chosen and does not distinguish the two structures for the activated complex. The experimental isotope effects for C_2D_4, CH_2CD_2 and *cis*-CHDCHD with the same geometrical structure and set of force constants were reproduced in a self-consistent manner for either models. Nonetheless, these calculations were useful.

Streitwieser and Fahey (*16*) made the first observation of a secondary α-H/D kinetic isotope effect on an S_N1 solvolysis reaction in which the transition state is a carbonium ion with the positively charged carbon atom having an sp^2 hybridization. The isotope effect resulted (*17*) from the change in the frequencies of the out of plane CH bending modes during rehybridization of the carbon atom from the sp^3 to sp^2 hybrid state as the reaction progresses from reactants to activated complex. This interpretation of secondary α-effects has since been accepted and applied to other types of reactions, among them additions of atoms and radicals to olefins. We find from detailed calculations (*18*) that this interpretation of secondary α-H/D kinetic isotope effects does not hold for the $S + C_2H_4$ reaction and generally for the addition of atoms and radicals to olefinic double bonds.

To evaluate the increment to the isotope effect caused by each vibrational motion in the $S + C_2H_4$ reaction, the magnitude of the isotope effect associated with each normal mode was calculated separately. The results obtained for the asymmetrical model of the C_2D_4/C_2H_4 reactant pair are collected in Table II. The most important contribution to the isotope effect comes not from the out-of-plane CH bendings but from a single vibration—namely the asymmetric twist mode of the thiirane molecule which is absent in the ethylene.

These results are valid and apply for all addition reactions involving olefinic double bonds. Addition reactions are characterized by an increase in the number of normal modes of vibration. In this case the ethylene molecule has 12 normal modes of vibration while thiirane has 15. One of these, the CS stretching mode, coincides with the reaction coordinate and does not contribute to the isotope effect. Out of the net gain of two, the CCS bending mode is not sensitive to isotopic substitution and does not generate an isotope effect, but the twist of the CH_2 group which

Table II. Calculated Increments of Individual Vibrational Motion to the Kinetic Isotope Effect for the Asymmetrical Model of the C_2D_4/C_2H_4 Reactant Pair[a]

	Vibration, cm^{-1}			
	H=H	H—H...S	$(EXC)^{-1}$	$(ZPE)^{-1}$	$(EXC \times ZPE)^{-1}$
CC stretch	1623	975	1.006	0.980	0.986
CH$_2$ rock	1236 / 810	1236 / 750	1.00 / 1.012	1.00 / 0.986	1.00 / 0.998
CH$_2$ deform	1342 / 1442	1450 / 1443	0.991 / 1.00	1.032 / 1.00	1.023 / 1.00
CH$_2$ wag	43 / 949	1260 / 949	0.99 / 1.00	1.26 / 1.00	1.25 / 1.00
CH$_2$ twist	1027	772 / 1310	1.012 / 1.022	0.853 / 1.78	0.864 / 1.81

$\Pi(EXC \times ZPE)^{-1} = 1.97$; $k(C_2D_4)/k(C_2H_4) = (MMI \times EXC \times ZPE)^{-1} = 1.28$.

[a] The transition state is assumed to be identical with the final product, the $\cdot CH_2CH_2S \cdot$ diradical.

constitutes the reaction center towards the CCS plane is isotope sensitive and is the main source of the isotope effect.

The experimental and calculated values of isotope effects are nearly independent of temperature in the range 28°–158°C and are a result of changes in the A-factor of the reaction. Model calculations on hypothetical, heavier substituents indicate an increase in the A-factor with increasing mass. The magnitude of this increase accounts for the observed increase in the A-factor of the halogenated ethylenes and explains a number of similar phenomena. The effect is referred to as a generalized secondary α kinetic isotope effect.

A theoretical study of the interaction of sulfur atoms with ethylene within the framework of the Extended Hückel MO theory has been reported by Hoffmann and co-workers (19). Potential surface calculations revealed two minima for the $S(^1D_2) + C_2H_4$ system. The higher corresponds to vinyl mercaptan formation via C–H bond insertion, and the lower, lying about 20 kcal below the former, to the least-motion, symmetry-allowed addition of sulfur across the double bond. The two are viewed as competing concerted processes. Similar calculations for the

$S(^3P)$ configuration indicated one minimum in the potential surface, leading to thiirane formation. The stereospecificity of the addition reaction was attributed to a correlation with an excited state of thiirane (*19, 20*) which retains CC bonding but which is unstable with respect to CS ring opening. The ring-opened thiirane intermediate has a CCS bond angle of 110°, and the terminal methylene group plane is perpendicular to the CCS plane. The calculated energy barrier for rotation of the methylene was 5 kcal.

More reliable Hartree–Fock type *MO* calculations (*21*) using Gaussian type wavefunctions built up by contraction from primitive Gaussian type orbitals (up to a 48 spd basis set) yielded results which are in general agreement with Hoffmann *et al.*'s EHMO results (*19*). Singlet and triplet excited configurations of thiirane were constructed from the ground electronic state MO wavefunctions by the virtual orbital method. Potential energy profiles for the ground state and the lowest excited vertical singlet and triplet states are shown in Figure 1. The ring-distorted struc-

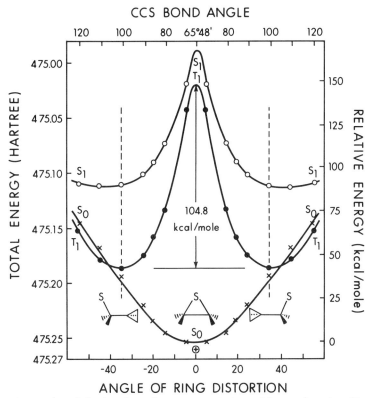

Figure 1.   Calculated total energy variation with angular ring distortion for the ground state (S$_0$), lowest excited triplet state (T$_1$), and lowest excited singlet state (S$_1$) of the thiirane molecule

Table III. Arrhenius Parameters for the Addition of
S(^3P_J) Atoms to Acetylenes[a]

	$\dfrac{E_a(C_2H_4) - E_a}{kcal/mole}$	$A/A(C_2H_4)$
C_2H_2	-2.0	6.2
$CH_3C\equiv CH$	-0.9	6.2
$CH_3C\equiv CCH_3$	1.3	2.7

[a] Temperature range: 25–ca. 125°C.

tures with a CCS angle of $\sim$100° represent the stable configurations of
these excited states, and transitions of the vertical states to the ring-
distorted structure are downhill processes. Therefore, even if the vertical
triplet state of the thiirane is reached, its rate of transition to the ring-
distorted structure is rapid, $k \sim 10^{13}$ sec^{-1}, and a mechanism for triplet
sulfur atom addition should include the ring-distorted thiirane as an
intermediate of the reaction.

It is required that this ring-distorted, triplet state thiirane intermediate possess a substantial energy barrier with respect to rotation about the
CC bond to maintain the stereochemical information content of the reaction. Calculations yielded a value of 23.6 kcal for the rotational energy
barrier, which explains the experimental results.

The homogeneous, gas phase, thermal decomposition of thiiranes,
ethylene episulfide, propylene episulfide, and 2-butene episulfide sheds
additional light on the mechanism of the S + olefin reactions. Below
ca. 250°C the decomposition products are sulfur and the olefin. The rate
of olefin formation is first order in episulfide concentration, and the reaction features an activation energy which is considerably lower than the
endothermicity of the expected unimolecular reaction $\diagdown\diagup \to C_2H_4 +$
S
S(^3P). The kinetics are consistent with a reaction mechanism involving
an electronically excited thiirane intermediate, and the experimental ac-

Table IV. Rate Constants and Activation Energies for the

	k (liter mole^{-1} sec^{-1})[a]	
	$25°C$	$80°C$
Ethylene	$(1.25 \pm 0.3) \times 10^7$ (12)	$(2.4 \pm 0.5) \times 10^7$ (11)
Propylene	$(1.18 \pm 0.27) \times 10^8$ (11)	$(1.37 \pm 0.27) \times 10^8$ (7)
Butene-1	$(1.5 \pm 0.4) \times 10^8$ (4)	—
cis-Butene-2	$(6.3 \pm 0.8) \times 10^8$ (5)	—
Tetramethyl- ethylene	$(3.85 \pm 0.8) \times 10^9$ (6)	$(2.6 \pm 0.9) \times 10^9$ (6)

[a] Numbers in parenthesis indicate number of measurements.

tivation energies $\sim$40, 39, and 36 kcal for C_2H_4S, C_3H_6S, and C_4H_8S, respectively, are identified with the excitation energies of the lowest triplet states of the thiirane molecules. These values lie near those predicted by the EHMO and the *ab initio* calculations for the lowest non-vertical triplet state of thiirane.

The reactions of $^1D_2$ sulfur atoms, with acetylenes afford the unstable thiirene (22), the family of unsaturated episulfides:

These adducts have been detected in flash photolysis experiments by kinetic mass spectrometry and shown to have half-lives a tenth to several seconds at room temperature. They react readily with acetylenes to give thiophenes

$$ \begin{array}{c} \diagup \\ C{=}C \\ \diagdown \diagup \\ S \end{array} \quad + \text{RC}{\equiv}\text{CR} \rightarrow \begin{array}{c} \diagup R \\ \square \; \square \\ \diagdown \diagup R \\ S \end{array} $$

and solid polymeric materials as end products. The yields of thiophene with acetylene, propyne, and 2-butyne are 5–15% but with perfluoro-2-butyne it is >50%.

Another convenient source of thiirenes is the photolysis of 1,2,3-thiadiazoles,

$$ \begin{array}{c} \text{CH}{-}\text{N} \\ \| \qquad \| \\ \text{CH} \qquad \text{N} \\ \diagdown \diagup \\ S \end{array} \quad + h\nu \rightarrow \begin{array}{c} \text{CH}{=}\text{CH} \\ \diagdown \diagup \\ S \end{array} \quad + N_2 $$

Addition of $Te(^3P_2)$ Atoms to Olefins

E_a *(kcal mole^{-1})*	log A *(1 mole^{-1} sec^{-1})*
2.5 ± 1.0	8.93 ± 0.65
0.6	8.5
—	—
—	—
-1.6 ± 1.4	8.41 ± 0.94

In the presence of an acetylene, thiophene is produced. The addition occurs across the C–S bond of the thiirene, producing 2,3-substituted thiophenes. Photolysis of 4-methyl or 5-methyl thiadiazole yields only one thiophene, methylated in the 5-position, indicating a preferential addition on the unsubstituted side of the thiirene ring:

In contrast to these results oxirenes, which are intermediates of the photochemical Wolff rearrangements (23–25) of diazoketones, cannot be trapped with simple olefins or acetylenes.

The $S(^3P)$ atom addition reactions with acetylene feature higher activation energies and much higher A-factors than the olefin reactions (Table III). The trends with methyl substituents, however, remain; the activation energy decreases, and the A-factor falls off, as with other electrophilic reagents. Szwarc and co-workers have attributed the large A-factors of the acetylene reactions to a gain in the entropy of activation, as compared with the olefin reactions, in going from the linear acetylene to the nonlinear activated complex.

A reference to the similarity of the chemical behavior of the group VIA elements is appropriate. This similarity became evident since not only oxygen, sulfur, and selenium atoms react with unsaturated hydrocarbons but also tellurium atoms. $Te(^3P_2)$ atoms from the flash photolysis of $Te(CH_3)_2$ have been shown (6, 26) to form an unstable adduct with olefins. The rate constants and Arrhenius parameters obtained for a few olefins by kinetic absorption spectroscopy are tabulated in Table IV. The activation energies and the pre-exponential factors are of about equal magnitude as for the other elements of the group; the A-factors are one order of magnitude larger, and the activation energies are slightly larger.

Acknowledgments

I thank all my former and present associates who have contributed to the research covered in this article. Particular appreciation is due to H. E. Gunning and I. G. Csizmadia for stimulating discussions.

Literature Cited

(1) Donovan, R. J., Kirsch, L. J., Husain, D., *Nature* (1969) **222**, 1164.
(2) Sidhu, K. S., Lown, E. M., Strausz, O. P., Gunning, H. E., *J. Amer. Chem. Soc.* (1966) **88**, 254.

(3) Strausz, O. P., "Organosulfur Chemistry," pp. 11–32, M. J. Janssen, Ed., Interscience, New York, 1967.

(4) Strausz, O. P., Wiebe, H. A., Gunning, H. E., to be published.

(5) Strausz, O. P., O'Callaghan, W. B., Lown, E. M., Gunning, H. E., *J. Amer. Chem. Soc.* (1971) **93,** 559.

(6) Connor, J., Van Roodselaar, A., Fair, R. W., Strausz, O. P., *J. Amer. Chem. Soc.* (1971) **93,** 560.

(7) Donovan, R. J., Husain, D., Fair, R. W., Strausz, O. P., Gunning, H. E., *Trans. Faraday Soc.* (1970) **66,** 1635.

(8) Cvetanovic, R. J., *Advan. Photochem.* (1963) **1,** 115.

(9) Pearson, J. M., Szwarc, M., *Trans. Faraday Soc.* (1964) **60,** 553.

(10) Owen, G. E., Pearson, J. M., Szwarc, M., *Trans. Faraday Soc.* (1965) **61,** 1722.

(11) Elias, L., *J. Chem. Phys.* (1963) **38,** 989.

(12) Callear, A. B., Tyerman, W. J. R., *Trans. Faraday Soc.* (1966) **62,** 371.

(13) Braun, W., Natl. Bureau of Standards, Washington, D. C., private communication, 1970.

(14) Bigeleisen, J., Wolfsberg, M., *Advan. Chem. Phys.* (1958) **1,** 15.

(15) Wolfsberg, M., Stern, M. J., *Pure Appl. Chem.* (1964) **8,** 225.

(16) Streitwieser, A., Jr., Fahey, R. C., *Chem. Ind. (London)* (1957) 1417.

(17) Streitwieser, A., Jr., Jagon, R. H., Fahey, R. C., Suzuki, S., *J. Amer. Chem. Soc.* (1958) **80,** 2326.

(18) Strausz, O. P., Safarik, I., O'Callaghan, W. B., Gunning, H. E., *J. Amer. Chem. Soc.* (February, 1972).

(19) Hoffmann, R., Wan, C. C., Neagu, V., *Mol. Phys.* (1970) **19,** 113.

(20) Leppin, E., Gollnick, K., *Tetrahedron Lett.* (1969) 3819.

(21) Strausz, O. P., Gunning, H. E., Denes, A. S., Csizmadia, I. G., *J. Amer. Chem. Soc.,* to be published.

(22) Strausz, O. P., Font, J., Dedio, E. L., Kebarle, P., Gunning, H. E., *J. Amer. Chem. Soc.* (1967) **89,** 4805.

(23) Csizmadia, I. G., Font, J., Strausz, O. P., *J. Amer. Chem. Soc.* (1968) **90,** 7360.

(24) Thornton, D., Gosavi, R. K., Strausz, O. P., *J. Amer. Chem. Soc.* (1970) **92,** 1768.

(25) Frater, G., Strausz, O. P., *J. Amer. Chem. Soc.* (1970) **92,** 6654.

(26) Connor, J., Greig, G., Strausz, O. P., *J. Amer. Chem. Soc.* (1969) **91,** 5695.

RECEIVED March 4, 1971.

11

Synthesis, Structures, and Bonding in Certain Sulfur and Selenium Chlorides and Organochlorides

KENNETH J. WYNNE

University of Georgia, Athens, Ga. 30601

Sulfur and selenium chlorides and organochlorides display a wide range of structural types and frequently phase dependent structural variations. Valence bond theory is the commonly utilized qualitative approach to bonding in these compounds. An alternative view based on qualitative three-center, four electron molecular orbital theory and differential orbital utilization has been developed in this paper. This approach is useful in rationalizing the solid state structure and acceptor behavior of sulfur(IV) and selenium(IV) chlorides and organochlorides. Divalent and tetravalent sulfur and selenium chlorides and organochlorides are prepared generally in a similar fashion but display contrasting stability and Lewis acid behavior. The $SeCl_2$ molecule, previously known only in the gas phase, is stabilized in the crystalline tetramethylthiourea adduct $SeCl_2(tmtu)$.

This paper deals with synthesis, structure, and bonding in tetravalent sulfur and selenium chlorides and organochlorides and sulfur and selenium dichloride. The similarities and differences which exist between the sulfur compounds and their selenium analogs are discussed. Tetravalent sulfur and selenium chlorides and organochlorides are covered first while SCl_2 and $SeCl_2$ are covered later.

Sulfur and Selenium Tetrachlorides and their Organo Derivatives

Synthesis and General Properties. Sulfur tetrachloride, a pale yellow crystalline solid, is prepared best by chlorination of sulfur dichloride at $-78°$ (1). Selenium tetrachloride also pale yellow is available com-

150

mercially but may be prepared by the direct chlorination of the element
(*2, 3, 4, 5*). Organosulfur trichlorides may be prepared (*6*) by the reaction of disulfides either with chlorine (Equation 1)

$$R_2S_2 + 3Cl_2 \rightarrow 2RSCl_3 \tag{1}$$

or more conveniently with sulfuryl chloride. Organoselenium trichlorides
may be prepared similarly by chlorination of diselenides (*7*) or by treatment of alkyl- or arylseleninic acids with concentrated HCl (*8*). Diorganosulfur and -selenium dichlorides are prepared by the reaction of an
appropriate diorganochalconide with chlorine (*9, 10, 11*) while triorganosulfonium and -selenonium salts may be prepared by the conventional
methods used to produce ammonium salts (*10, 12*).

Since their stability does not depend on the oxidizing ability of
chlorine, triorganosulfonium and -selenonium chlorides are stable as are
salts containing many other anions. However sulfur tetrachloride and its
mono- and diorgano derivatives are less stable thermally than their selenium analogs (Table I). For this reason few organo- and diorgano-
sulfur(IV) chlorides have been isolated, but these compounds have been
postulated frequently as reaction intermediates (*13, 14*). Important information concerning the structure and bonding in these reactive sulfur
intermediates is inferred from a study of their selenium counterparts,
which may be studied at temperatures where the sulfur analogs are
unstable.

Although $SeCl_4$ and its organo derivatives are more stable thermally
than the corresponding sulfur compounds, they share with the latter a
general sensitivity toward moisture and are decomposed completely upon
dissolution in water. In addition certain selenium compounds, particularly
dialkylselenium dihalides are light sensitive.

Structure. Most of the compounds discussed here are only stable or
are most stable as solids. It is, therefore, important to know the nature
of the interactions which lead to the enhanced stability in the solid state.

One way of viewing the structures of chalcogen halides and their
organo derivatives in the solid state starts with a consideration of the
coordination number (C.N.) of the central atom. This raises immediately
a serious question concerning what constitutes bonded or coordinated
atoms because varying chalcogen–halogen interactions at less than the

Table I.   Decomposition Points in the Solid State (°C)

SCl_4 (*1*)	−31	$SeCl_4$ (*4*)	196, sub
CH_3SCl_3 (*6*)	30	CH_3SeCl_3 (*7*)	81
$C_6H_5SCl_3$ (*6*)	<10	$C_6H_5SeCl_3$ (*8*)	133
$(p\text{-}Cl\text{-}C_6H_4)_2SCl_2$ (*9*)	0	—	
—		$(CH_3)_2SeCl_2$ (*11*)	61

van der Waals distances are sometimes encountered. For this discussion we are considering atoms closer than half the difference between the sum of the van der Waals radii and the sum of the single bond convalent radii to be coordinated. Thus arbitrarily we consider any S–Cl or Se–Cl interactions closer than *ca.* 2.84 and 2.98 A, respectively, to represent "coordination." Pauling's van der Waals and covalent radii (*15*) have been utilized in these calculations. Using this definition the C.N.'s for sulfur and selenium tetrachloride and organo derivatives are presented in Table II. These results are based on structural information and on some estimates made in the section below on bonding.

Few x-ray crystallographic studies have been carried out in this area, but a number of structures have been deduced from spectroscopic stud-

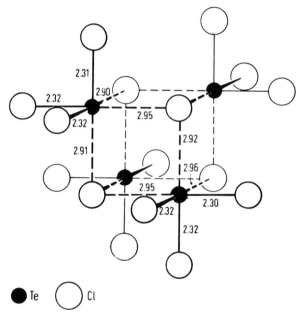

Angewandte Chemie International Edition in English

Figure 1.   Solid tellurium tetrachloride (18)

ies. The ionic formulation of $SeCl_3^+Cl^-$ has been proposed for $SeCl_4$ in the solid state on the basis of infrared (*16*) and Raman (*17*) spectroscopic studies. However, compounds apparently containing the $SeCl_3^+$ cation such as $SeCl_3^+SbCl_6^-$ displaying stretching frequencies (ν_1 and ν_3) for the $SeCl_3^+$ moiety which are about 50 cm^{-1} higher, respectively, than those found for the suggested $SeCl_3^+$ species in $SeCl_4$. Therefore most workers (*16*) have recognized that there is considerable cation–anion interaction

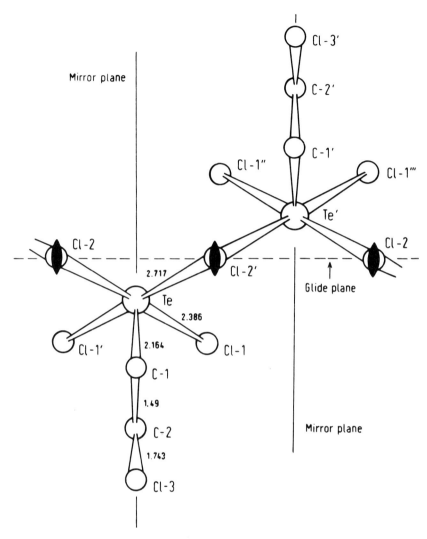

Figure 2.   *2-Chloroethyltellurium trichloride* (20)

**Table II.   Coordination Number for S and Se Tetrachloride
and Their Organo Derivatives**

S; C.N.	*Se; C.N.*	*Compound Class*
?	(6)?	MCl_4
(5)?	(5)	$RMCl_3$
4	4	R_2MCl_2
3	3	R_3MCl

(*i.e.* Se–Cl–Se halogen bridging) in SeCl$_4$. A detailed knowledge of the extent of this interaction will be known only when a complete x-ray crystallographic analysis has been performed. A guess is that the structure is found similar to that of tellurium tetrachloride (*18*) shown in Figure 1 with appropriate modifications discussed below in the section on bonding. Tellurium exhibits a C.N. of six in solid TeCl$_4$, having three close (2.32 A) and three more distant (2.92 A) Te–Cl interactions in isolated Te$_4$Cl$_6$ units. The limited spectral data available (*19*) do not permit an unambiguous choice of structure for SCl$_4$, but it may be "semi-ionic" as discussed below.

No detailed structural data are presently available for organosulfur or organoselenium trichlorides, but again a recently determined structure of a tellurium compound provides interest. The structure of β-chloroethyl-tellurium trichloride (Figure 2) shows tellurium to be five coordinate with two short terminal (2.39) and two longer bridging (2.72) Te–Cl bonds (*20*). Infrared spectral data (*7, 21*) suggest that Se in methylselenium trichloride also may be pentacoordinate. This may be achieved by linking RSeCl$_2$ units together through chlorine bridges in a manner analogous to that found for ClC$_2$H$_4$TeCl$_3$. Square pyramidal coordination would result about Se. Supporting the view that Cl bridging is important is the fact that the highest Se–Cl stretching frequency (v_2) in CH$_3$SeCl$_2^+$ (*21*) is 84 cm^{-1} higher than the highest Se–Cl stretching frequency in CH$_3$SeCl$_3$.

Considerable structural data is available concerning diorganosulfur and -selenium chlorides. Baenziger has reported the structure of bis(*p*-

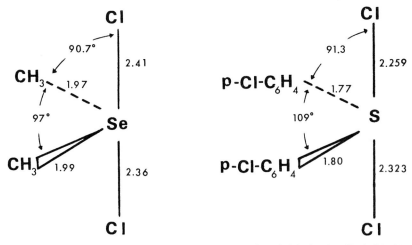

Journal of the American Chemical Society

Figure 3.   (CH$_3$)$_2$SeCl$_2$ (*24*) and (p–Cl–C$_6$H$_4$)$_2$SCl$_2$ (*9*)

chlorophenyl) sulfur dichloride (9). For selenium the structures of $(C_6H_5)_2SeCl_2$ (22), (p-CH_3-$C_6H_4)_2SeCl_2$ (23), and $(CH_3)_2SeCl_2$ (24) are known. The structures of (p-Cl-$C_6H_4)_2SCl_2$ and $(CH_3)_2SeCl_2$ shown in Figure 3 demonstrate that the central atom has a C.N. of four in these compounds. One of the salient features common to these molecules and others of this type is the long chalcogen–chlorine distance compared with the expected normal covalent radii sum (2.03 A for S–Cl, 2.16 A for Se–Cl).

No x-ray crystallographic data are available for a triorganosulfonium chloride, but $[(CH_3)_3S]^+I^-$ (25) has been shown to be ionic, as has the

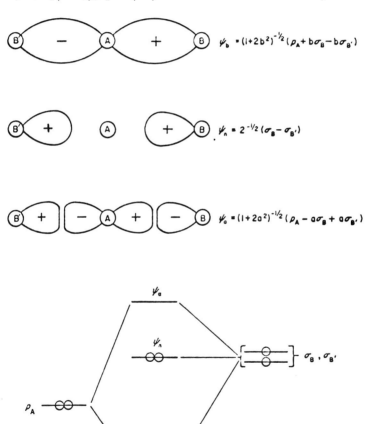

F. Cotton, G. Wilkinson, "Advanced Inorganic Chemistry," Interscience.

Figure 4.   (Upper) The formation of a set of three-center orbitals from a p-orbital on the central atom A and orbitals on the two outer B atoms. (Lower) An approximate energy level diagram (29).

selenonium salt $[(C_6H_5)_3Se]^+Cl^-$ (16). Infrared studies and general
physical properties have shown also that triorganosulfonium (27) and
triorganoselenium (7) halides are essentially ionic.

Bonding. Full valence hybridization, or valence bond theory is the
normally utilized bonding approach for organo derivatives of sulfur and
selenium tetrachloride and related compounds (28, 29). However, a delocalized bonding scheme has been proposed (29, 30, 31, 32) which
should be directly applicable to these compounds. It is worthwhile to
explore the applicability of the approach in order to examine whether
insight may be gained in rationalizing various structural trends, reactivities, and Lewis acceptor behavior of sulfur and selenium tetrachloride
and organo derivatives.

The bonding theory utilized here is "three center bond theory." This
is a limited MO treatment in which bonding is assumed to occur only
through valence p-orbitals. The formation of three center orbitals is
illustrated in Figure 4 together with an energy level diagram for a typical
case.

One aspect of the three-center MO bonding theory which appears
not to have been explored in detail previously is the consequence of differential orbital utilization. Consider a linear three center system C A C'
where, for whatever reason, C is bound more closely to A than C'. The
two electrons in Ψ_b will tend to associate more with the AC bond in
proportion to the differential binding of C and C'. Conversely, the pair
of electrons in Ψ_n will tend to become associated with more weakly bound
C'. Thus as the binding of C becomes greater the AC bond length
decreases while the AC' bond length increases. An example is found
in tellurium chemistry. Figure 5A shows the structure of *trans*-dibromo-
bis(ethylenethiourea)tellurium(II) (33), while Figure 5B shows that of
phenylbis(tetramethylthiourea)tellurium(II) chloride (34). [Preliminary
evidence (35) indicates that dichlorobis(tetramethylthiourea)tellurium-
(II) exists as the trans isomer.] The structures in Figure 5 and related
data indicate that substitution of halide in A (Figure 5) by a phenyl
group results in a drastic lengthening of the bond trans to carbon. This
is viewed as a result of efficient overlap of the tellurium $5p$ orbital by the
sp^2 orbital of carbon. This example shows essentially complete orbital
utilization by carbon leading to a two-center two electron Te–C bond
and no bond trans to carbon.

Intermediate cases of differential orbital utilization are known as well.
Thus in comparing structure A, Figure 5, with that of *cis*-dibromobis(thiourea)tellurium(II), Figure 5C (36), apparently sulfur in the thiourea
ligand is bound more strongly than bromine. The ability of one ligand
which binds more strongly to the central atom to decrease bonding trans
to itself has been called the "trans-bond-lengthening effect" (34, 36), but

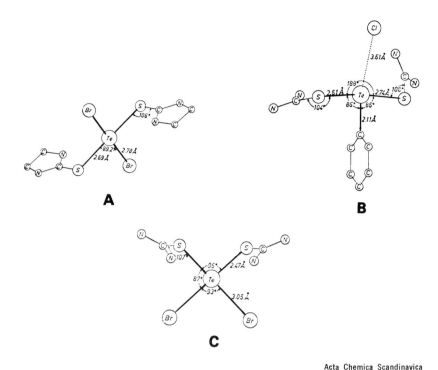

Figure 5.   (A) trans-*dibromobis(ethylenethiourea)tellurium(II)*, (B) phenyl-
bis(*tetramethylthiourea)tellurium(II)*, and (C) cis-*dibromobis(thiourea)tellu-
rium(II)* (33, 34, 36)

might be called "trans influence" in keeping with the normally accepted
definition of the term (*37*).

The postulation of a similar effect operating in sulfur(IV) and
selenium(IV) halides and organo derivatives helps rationalize the available structural data summarized in the previous section and allows some
predictions. Thus the strong trans lengthening effect of three R groups
connected to S or Se in triorganosulfonium or -selenonium compounds,
respectively, eliminates coordination trans to these groups with three-
coordinate R_3M^+ ions resulting. Coordinating groups ideally occupy three
sites of an octahedron cis to one another as shown in Figure 6A. Substitution of Cl for R opens up the possibility of bonding trans to one Cl leading
to four coordinate, neutral R_2MCl_2 species in which the groups occupy
four octahedral sites (Figure 6B). Substitution of one more R-group by
Cl opens up one more position for coordination trans to this chlorine. For
a neutral compound, however, utilization of this coordination site necessi-
tates the sharing of chlorines through Cl-bridging. The arrangement

which we believe results for RMCl₃ compounds in the solid state is shown in Figure 6C. The central atom has a C.N. of 5 with the groups in a distorted square pyramidal arrangement as a result of longer bridging M–Cl bonds. Finally, substitution of one more chlorine leads to a maximum utilization of coordination sites with a predicted coordination number of 6 for Se in SeCl₄ (and SCl₄?).

Support for the view on structure and bonding comes first from agreement with observed structural variations presented in the previous section. This approach allows a rationalization of why the drastic change from ionic to molecular structures occurs in going from triorganosulfonium and -selenonium chlorides to diorganosulfur and -selenium dichlorides. This theory is consonant with evidence of associative behavior for CH_3SeCl_3 (7) and allows the speculation that $RSCl_3$ compounds contain pentacoordinate sulfur, achieved in the solid state through Cl-bridging.

This approach is also consistent with bond lengths and angles observed in organosulfur and -selenium chlorides. Thus longer (~0.2 A greater than sum of single bond covalent radii) S–Cl and Se–Cl bond lengths are observed (9, 24) in compounds containing three center Cl–M–Cl bonds where the predicted bond order is 0.5.

The theory proposed here is especially useful because maximum coordination (resulting in a sort of intermolecular base stabilization) is attained generally in the solid state for the compounds under discussion.

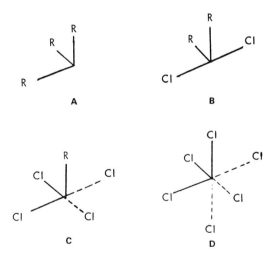

Figure 6. Idealized structural models for S and Se (M) tetrachlorides and organo derivatives in the solid state: A, R_3M^+; B, R_2MCl_2; C, $RMCl_3$; D, MCl_4

As a result, the approach presented here has an advantage over the full valence hybridization [or electron pair repulsion theory (38)] view. The latter theory does not rationalize readily the structure of solid TeCl₄ though it is successful with gaseous TeCl₄.

We next consider the relative importance of the single M–Cl bond *vs.* the three center Cl–M–Cl bond. As the central p orbital–Cl orbital overlap increases, the stability of a two center bond relative to the three center bond should increase. Therefore the overlap of the central p orbital per Cl orbital presumably would be greater for a single bond than for a symmetrical multicenter bond where the S–Cl or Se–Cl bond order would theoretically be 0.5. Sulfur should offer the best energy match and greater overlap for a Cl p orbital so that a singly bonded S–Cl system should be more stable than one containing a Cl–S–Cl group. This is found in the instability of SCl_4, CH_3SCl_3, and $(CH_3)_2SCl_2$ where three-center bonds occur as contrasted with the stability of the SCl_3^+, $CH_3SCl_2^+$, and $(CH_3)_2SCl^+$ cations (27) where only two center S–Cl bonds would exist. These ideas are consonant with the marked increase in the stability of MCl_4 compounds and their organo derivatives in proceeding from sulfur to selenium and the comparable stability of compounds containing MCl_3^+, $RMCl_2^+$, or R_2MCl^+ cations (11, 21, 27).

The increasing importance of single bonds in going from tellurium to sulfur may result in increased differential orbital utilization in cases where halogen bridging occurs (RMX_3, MX_4). This together with the smaller size of sulfur could increase the ionic nature of the compounds substantially.

Finally, the three-center bond MO theory together with the differential orbital utilization concept prove valuable in rationalizing acceptor behavior of organotellurium trihalides (39). Direct extension of the theory to sulfur and selenium analogs is possible (though space limitations does not permit this here) and provides structural predictions for possible adducts and intermediates.

To a degree the above theory rationalizes the range of structures of organochalcogen(IV) and chalcogen(IV) halides. Some of the limitations of this approach are given now. A general criticism of the three-center bond theory has been presented (29). The utility of present theory which combines the three center bond theory with the "trans bond-lengthening effect" is based on the large difference in orbital utilization between alkyl and aryl groups and the halogens. This leads to a fairly clear distinction between coordinated and noncoordinated moieties. Although it seems that organo groups have a strong trans bond-lengthening effect, limited interaction may take place trans to these groups (9, 40, 41). In situations where differences in trans bond-lengthening influence are smaller, structures would be more complex.

Sulfur and Selenium Dihalides

We move now to some selected aspects of the chemistry of sulfur and selenium dichloride. Sulfur dichloride is an important chemical which is utilized in rubber vulcanization, as a chlorinating agent, as a catalyst in many systems, and in other ways. Sulfur dichloride is prepared from S_2Cl_2 using excess chlorine and a suitable catalyst. It is isolated by fractional distillation in the presence of PCl_5 which stabilizes the product (42). Sulfur dichloride is a reactive substance and most of the reported chemistry of this compound concerns substitution or redox reactions. However, a note reports the formation of an adduct of the composition $SCl_2(py)_2$ with pyridine (43). $SCl_2(py)_2$ is a white crystalline material which melts at 97°–98°C. From these results it appears that the base stabilization of SCl_2 simultaneously results in the formation of easily handled crystalline material. This area is one that should be examined more thoroughly with a view toward obtaining detailed information concerning the relative reactivity of various adducts, the degree of stabilization of SCl_2, and the structure of these novel compounds.

In contrast to SCl_2 which, though reactive, is metastable in the liquid phase, $SeCl_2$ has been known previously only in the gas phase in the presence of chlorine (44, 45). Until our current investigation the only known chemistry of $SeCl_2$ was the disproportionation to Se_2Cl_2 and $SeCl_4$ in the liquid or solid states and the reaction with Cl_2 to give $SeCl_4$.

We have reported recently the preparation of methyltrihalo(tetramethylthiourea)tellurium(IV) complexes (39) from the reaction of CH_3TeCl_3 and CH_3TeBr_3 with tetramethylthiourea(tmtu). Attempting to extend our studies we found that the reaction of methylselenium trichloride with tmtu gave the unexpected complex $SeCl_2(tmtu)$ according to Equation 2 (46).

$$CH_3SeCl_3 + tmtu \rightarrow CH_3Cl + SeCl_2(tmtu) \tag{2}$$

Dichloro(tetramethylthiourea)selenium(II) is a moisture sensitive, yellow crystalline compound which is recrystallized from methylene chloride or chloroform.

Cryoscopic molecular weight studies indicated the compound was monomeric in nitrobenzene solution. A proton nuclear magnetic resonance spectrum showed the tmtu resonance in $SeCl_2(tmtu)$ was shifted downfield by ∼0.3 ppm and gave further evidence that the adduct was undissociated and monomeric in solution.

Our infrared spectral data suggest no gross structural changes occur in going from solution to the solid state. Although neither our infrared nor other data provide direct information concerning the molecular structure, the T-shaped molecule shown below

$$[(CH_3)_2N]_2 \ \overset{\overset{\displaystyle Cl}{|}}{\underset{\underset{\displaystyle Cl}{|}}{C-S-Se}}$$

seems a reasonable structural model and is analogous to that found for
$TeCl_2(tmtu)$ (*48*). [We have recently determined the crystal and molecular structure of dibromo(tetramethylthiourea)selenium(II), $SeBr_2$-
(tmtu), and have found a T-shaped molecular structure in the solid state
analogous to that proposed above for $SeCl_2(tmtu)$.]

Our "trapping" of the $SeCl_2$ molecule suggests it is an important
intermediate in the decomposition of alkylselenium trichlorides. The
Lewis base stabilization of $SeCl_2$ is reminiscent of the stabilization of
S_2O molecule in $S_2O \cdot (CH_3)_3N$ (*49*).

Species analogous to $SeCl_2(tmtu)$ may be isolable in sulfur chemistry
and/or may be important reaction intermediates. Work is continuing on
the isolation and identification of these and related compounds.

Acknowledgment

The author thanks the National Science Foundation for generous
support of this research through Grant GP 9486 and Canadian Copper
Refiners Ltd. for a gift of selenium through the Selenium–Tellurium Development Association.

Literature Cited

(1) Ruff, O., *Chem. Ber.* (1904) **37**, 4513.
(2) Brauer, G., "Handbook of Preparative Inorganic Chemistry," Vol. 1, p.
    423, Academic Press, 1965.
(3) Nowak, H. G., Suttle, J. F., *Inorg. Syn.* (1957) **5**, 125.
(4) Simons, J. H., *J. Amer. Chem. Soc.* (1930) **52**, 3483.
(5) Yost, D. M., Kinrher, C. F., *J. Amer. Chem. Soc.* (1930) **52**, 4680.
(6) Douglass, I. B., Brower, K. R., Martin, F. T., *J. Amer. Chem. Soc.* (1952)
    74, 5770.
(7) Wynne, K. J., George, J. W., *J. Amer. Chem. Soc.* (1969) **91**, 1649.
(8) Foster, D. G., *J. Chem. Soc.* (1933), 822.
(9) Baenziger, N. C., Buckles, R. E., Mauer, R. J., Simpson, T. D., *J. Amer.
    Chem. Soc.* (1969) **91**, 5749.
(10) Campbell, T. W., Walker, H. G., Coppinger, J. M., *Chem. Rev.* (1952)
    50, 279.
(11) Wynne, K. J., George, J. W., *J. Amer. Chem. Soc.* (1965) **87**, 4750.
(12) Geordeler, J., "Methoden der Organischen Chemie," Vol. 9, p. 171, G. T.
    Verlag, Stuttgart, 1955.
(13) Schoberl, A., Wagner, A., *Ibid.*, p. 211.
(14) Szmant, H. H., "Organic Sulfur Compounds," p. 154, N. Kharasch, Ed.,
    Pergamon Press, New York, 1961.
(15) Pauling, L., "The Nature of the Chemical Bond," 3rd ed., Cornell Univer-
    sity Press, Ithaca, New York, 1960.
(16) George, J. W., Katsaros, N., Wynne, K. J., *Inorg. Chem.* (1967) **6**, 903.

(17) Gerding, H., Houtgraaf, H., *Rec. Trav. Chim. Pays Bas* (1954) **73**, 737.
(18) Buss, B., Krebs, B., *Angew. Chem. Int. Ed. Engl.* (1970) **9**, 463.
(19) Beattie, I. R., Chudzynska, H., *J. Chem. Soc. A* (1967) 984.
(20) Kobelt, D., Paulus, E. F., *Angew. Chem. Int. Ed. Engl.* (1971) **10**, 74.
(21) Wynne, K. J., Pearson, P. S., *Inorg. Chem.* (1971) **10**, 1871.
(22) McCullough, J. D., Hamburger, G., *J. Amer. Chem. Soc.* (1942) **64**, 508.
(23) McCullough, J. D., Marsh, R. E., *Acta. Crystallogr.* (1950) **3**, 41.
(24) Cordes, A. W., Symposium on Stereochemistry of Inorganic Compounds, Banff, Alberta, Canada (June 1968).
(25) Zuccaro, D. E., McCullough, J. D., *Z. Kristallogr.* (1959) **112**, 401.
(26) McCullough, J. D., Marsh, R. E., *J. Amer. Chem. Soc.* (1950) **72**, 4556.
(27) Peach, M. E., *Can. J. Chem.* (1969) **47**, 1675.
(28) Abrahams, S. C., *Quart. Rev.* (1956) **10**, 407.
(29) Cotton, F., Wilkinson, G., "Advanced Inorganic Chemistry," 2nd ed., Chapt. 15, 1966.
(30) Foss, O., *Acta. Chem. Scand.* (1962) **16**, 779.
(31) Musher, J. I., *Angew. Chem. Int. Ed. Engl.* (1969) **8**, 54.
(32) Musher, J. I., *Advan. Chem. Ser.* (1972) **110**, 44.
(33) Foss, O., Kjøge, H. M., Marøy, K., *Acta Chem. Scand.* (1965) **19**, 2349.
(34) Foss, O., Marøy, K., *Acta Chem. Scand.* (1966) **20**, 123.
(35) Foss, O., Johannesen, W., *Acta Chem. Scand.* (1961), 1941.
(36) Foss, O., Johnsen, K., Maartman-Moe, K., Marøy, K., *Acta Chem. Scand.* (1966) **20**, 113.
(37) Zumdahl, S. S., Drago, R. S., *J. Amer. Chem. Soc.* (1968) **90**, 6669.
(38) Gillespie, R. J., *J. Chem. Soc.* (1963), 4672.
(39) Wynne, K. J., Pearson, P. S., *Inorg. Chem.* (1971) **10**, 2735.
(40) Christofferson, G. D., Sparks, R. A., McCullough, J. D., *Acta Cryst.* (1958) **11**, 782.
(41) Hope, H., *Acta Cryst.* (1966) **20**, 610.
(42) Rossler, R. J., Whitt, F. R., *J. Appl. Chem.* (1960) **10**, 299.
(43) Wannagat, V., Schindler, G., *Angew. Chem.* (1957) **69**, 784.
(44) Ozin, G. A., Vander Voet, A., *Chem. Commun.* (1970), 896.
(45) Simons, J. H., *J. Amer. Chem. Soc.* (1930) **52**, 3483.
(46) Wynne, K. J., Pearson, P. S., *Chem. Commun.* (1971), 293.
(47) Pearson, P. S., Newton, M. G., Wynne, K. J., unpublished results.
(48) Foss, O., Johannesen, W., *Acta Chem. Scand.* (1961) **15**, 1940.
(49) Schenk, P. W., Steudel, W., *Angew. Chem. Int. Ed. Engl.* (1965) **4**, 402.

RECEIVED March 5, 1971.

12

Investigation of Amorphous Chalcogenide Alloys Using Laser Raman Speetroscopy

ANTHONY T. WARD

Research Laboratories, Xerox Corp., Rochester, New York

*Laser Raman spectroscopy has been used as a tool to eluci-
date the molecular structure of crystals, liquids, and amorphous alloys in the As–S–Se–Te system. Characteristic
monomer and polymer structures have been identified, and
their relative abundances have been estimated as a function
of temperature and atomic composition. These spectroscopic
estimates are supported by calculations based on the equilibrium polymerization theories of Tobolsky and Eisenberg
(1, 2) and of Tobolsky and Owen (3). Correlations between
the molecular structure of the amorphous alloys and physicochemical properties such as the electron drift mobility and
the glass transition temperature are presented.*

Xerox's interest in sulfur dates back to 1938 when Chester Carlson first demonstrated the phenomenon of electrophotography using a sulfur-coated zinc plate as the photoactive element of an imaging system (4). The principal steps of Carlson's process are used in present-day Xerox copiers. As shown in Figure 1 the process involves:

(1) Deposition of a uniform electric charge on the photoreceptor surface. (Carlson achieved this by rubbing the plate with a silk handkerchief.)

(2) Imagewise exposure to a light source causing discharge in the illuminated areas.

(3) Development of the electrostatic image by dusting with a pigmented toner carrying a charge of opposite sign. (Carlson used lycopodium.)

(4) Electrostatic transfer of the toned image to paper.

(5) Fixing of the image, usually by heat.

(6) Cleaning of the photoreceptor.

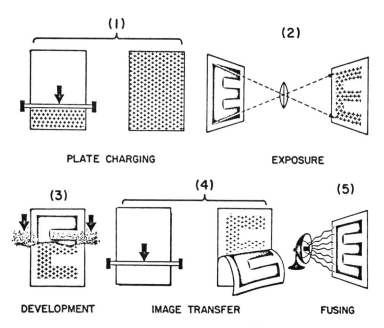

(1)   **(2)**

PLATE CHARGING   EXPOSURE

(3)   **(4)**   **(5)**

DEVELOPMENT   IMAGE TRANSFER   FUSING

Figure 1.   Basic steps in xerographic imaging

Subsequent development of the invention at Battelle Memorial Institute between 1944 and 1948 showed that the speed of the imaging device could be improved by alloying the sulfur with selenium. Eventually, the advantages of using pure vitreous selenium were recognized, and this material was used in the first machines to appear on the market in 1950. Since then much effort has been devoted to the development of a basic understanding of those characteristics of vitreous selenium which make it superior xerographically.

The ability of selenium to adopt a stable vitreous form was an important factor. Many of the initial basic studies were concerned with the unique physical properties of the amorphous state and its influence on optical, electrical, and thermodynamic parameters. An example of this influence appears in the work of Stuke (5) and of Hartke and Regensburger (6) (Figure 2). This diagram is a plot of optical absorption and photoresponse as a function of photon energy for (a) trigonal selenium and (b) vitreous selenium. This result suggests that even the properties of vitreous selenium could be improved. The vitreous material has a region of non-photoconductive absorption between 2.0 and 2.3 ev—*i.e.,* a spectral region for which the incident light cannot produce an image in the xerographic mode. This inefficiency in utilizing the incident light, if overcome, could lead to more efficient imaging and to higher copying speeds.

Concurrently with these physical investigations, interchalcogenide chemical relationships were being explored also, and these became important later in formulating new photoconductive binary and ternary alloys from elements in groups V and VI of the Periodic Table. Thermodynamic studies of these alloys indicated possible avenues of correlation between the physical properties of the amorphous state and the chemical properties of the alloy constituents. An example is shown in Figure 3 taken from the work of Myers and Felty (7). Here the liquidus temperature and glass transition temperature (T_g) determined by scanning differential calorimetry are plotted as a function of atomic composition for the arsenic–sulfur binary system. The linear increase in T_g between 10 and 40 atomic percent of arsenic was attributed to the existence, in the glassy phase, of a network of As and S atoms covalently bound to allow satisfaction of local valence. A qualitatively similar result was observed for the As–Se system: addition of arsenic to selenium increased the glass transition temperature and the stability of the vitreous phase by forming a crosslinked polymer network. This property was used in compact desktop copiers where the proximity of photoreceptor and fusing system required increased thermal stability of the photoconductor.

Such considerations emphasized the need to confirm the microscopic origins of physicochemical correlations, and thus, far-infrared and laser

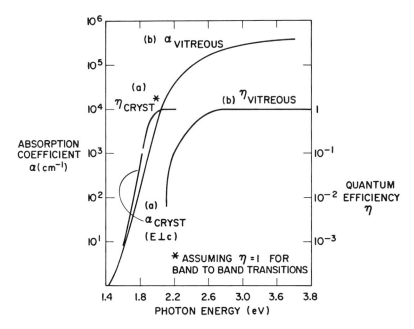

Figure 2.   Optical absorption and photoresponse as a function of photon energy for (a) trigonal selenium, (b) vitreous selenium

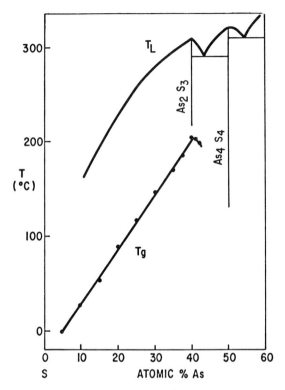

*Figure 3.   Glass transition temperatures and liquidus temperatures
for the As–S system*

Raman spectroscopy were activated. The results of the two techniques
proved complementary. Limits set by the title of this paper permit only
a description of the Raman activity.

The apparatus is shown schematically in Figure 4, configuration A
being used for crystals and B for liquids and glasses. The incident beam,
either 50–100 mwatts at 6328A from a Spectra physics model 125 He:Ne
laser or 10–15 watts at 1.06μ from a Korad KY-2 ND^{3+}:YAG laser, (depending on the transmission properties of the sample), is focused into the
sample, and the light scattered at 90° to the incident beam direction is
collected and analyzed by a Spex 1400 double monochromator. An
alternative configuration involving 45° incidence of the laser light and
normal collection of the scattered light is used for evaporated films. Under
these conditions, discrete lines or peaks appear in the scattered light
spectrum which result from inelastic scattering of the incident photons
by polarizability changes associated with the molecular vibrations of the
sample. These lines are shifted in energy from the incident photon energy

by amounts corresponding to the characteristic vibrational energies and constitute a unique "fingerprint" of the vibrating species present.

Typical Raman spectra are shown in Figure 5. They demonstrate the structural relationship between crystalline and disordered forms of sulfur and selenium. The disordered form of sulfur in this example is the liquid at the melting point and that for selenium is the vacuum-evaporated or melt-quenched vitreous material at room temperature. The correspondence of Raman line frequencies at 151, 218, and 474 cm^{-1} for the rhombic crystal and the liquid at 120°C confirms the identity of the S_8 molecule in both phases of sulfur. Similarly, in the case of vitreous selenium the occurrence of frequency components at 239 and 250 cm^{-1} characteristic of both crystalline modifications of selenium strongly suggests the coexistence of the Se_8 rings and the Se_n chains found in the respective crystals. However, the spectral broadening of the Raman feature at 239 cm^{-1} corresponding in frequency with the Se_n vibration frequency of trigonal selenium indicates that the helical axis of symmetry of the chains is lost in going from the crystal to the vitreous phase.

These results provide qualitative support for the validity of the ring-chain structural model for vitreous selenium implied by the Tobolsky–Eisenberg theory (*1, 2*) of equilibrium polymerization. The establishment

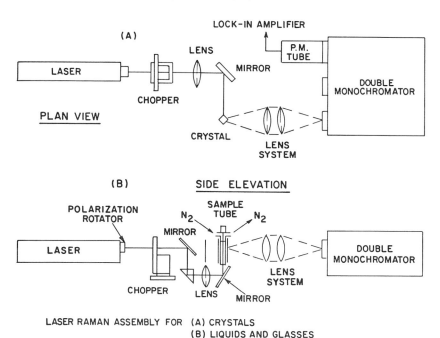

LASER RAMAN ASSEMBLY FOR (A) CRYSTALS
(B) LIQUIDS AND GLASSES

Figure 4.   Experimental configurations used for laser Raman spectroscopy of (A) crystals, (B) liquids and glasses

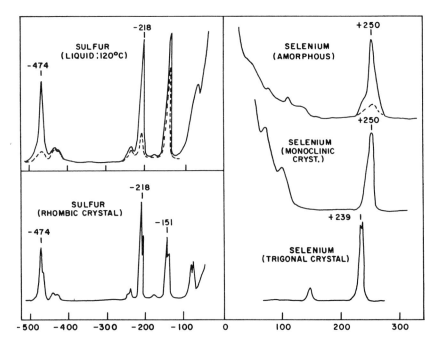

Figure 5. Raman spectra of crystalline and disordered forms of sulfur and selenium

of the general applicability of the equilibrium polymerization concept thus assumes particular importance because of its potential application to determining vitreous microstructure. Quantitative support for this theory is derived by observing the temperature dependence of the Raman spectrum of liquid sulfur. The equilibrium conversion of S_8 monomer rings to polymeric S_n chains which occurs at 159°C is followed conveniently as a function of temperature by monitoring the S_8 Raman line intensities as shown by Ward and Myers (8). The result, shown by the curves designated ϕ_M (sulfur) and ϕ_P (sulfur) in Figure 6, indicates the percentage of sulfur atoms present as S_8 monomer and S_n polymer, and agrees with the Tobolsky–Eisenberg prediction. The other pair of curves designated ϕ_M (arsenic–sulfur) and ϕ_P (arsenic–sulfur) indicates the corresponding result for arsenic–sulfur alloys containing 5 to 15 atomic % As. Again the form of the curves is rationalized by invoking the equilibrium polymerization concept. However in the arsenic–sulfur case, ϕ_M and ϕ_P are not absolute quantities. They represent changes in the concentrations of monomer and polymer relative to the concentrations quenched into the alloys during their initial preparation. Neither the Raman method nor the polymerization theory permit derivation of the absolute quantities in this particular case. Vitreous systems to which the theory is applied are discussed later. A qualitative idea of the initial monomer/polymer ratio in As–S alloys is gained from Figure 7 which shows the Raman spectra of these alloys as a function of arsenic concentration between 0 and 35 atomic % of As. The broad peak near 345 cm^{-1} is caused by a fundamental vibration of the As–S polymer network. The other lines constitute the familiar spectral signature of the S_8 species shown earlier. With increasing arsenic concentration the scattering contribution resulting from polymer increases at the expense of that attributed to S_8 monomer. This increase in the extent of network formation accounts for the linear increase in T_g observed previously in this composition range by Myers and Felty (7). Their work, shown in Figure 3,

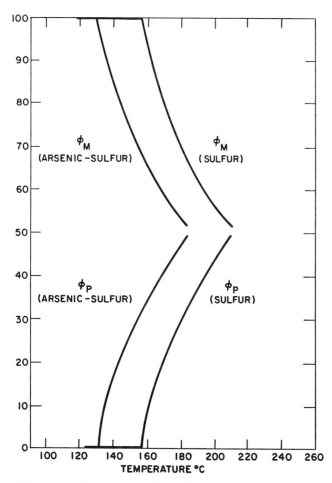

Figure 6.   Temperature dependence of the fraction of sulfur present as monomer and polymer in liquid sulfur and liquid arsenic–sulfur alloys

indicates that atomic concentrations of arsenic greater than 40% should produce breakdown of the polymer network, a decrease in T_g, and de-vitrification, caused by crystallization of realgar, As_4S_4, at the stoichio-metric composition $As_{0.50}$. These changes are followed spectroscopically as shown in Figure 8; the broad peak at 345 cm^{-1} caused by polymer decays while sharp lines characteristic of crystalline realgar appear as the arsenic concentration increases. There is some evidence in the form of peaks at 190 cm^{-1} and 220 cm^{-1} for the presence of As_4S_4 monomer in $As_{0.40}S_{0.60}$, a nominally stoichiometric composition. The relative intensity of these peaks, and therefore the concentration of As_4S_4, varies with the

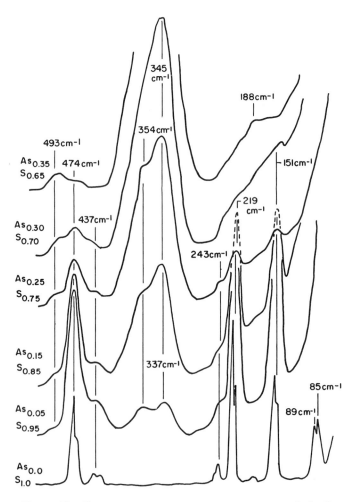

Figure 7. Raman spectra at room temperature of As–S alloys containing 0–35 atomic % As

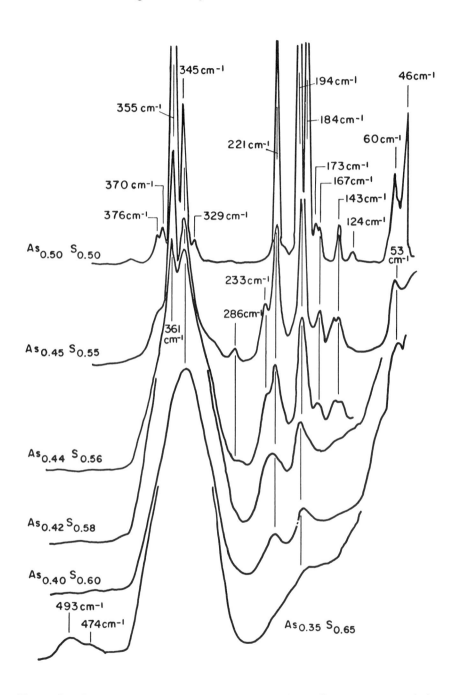

Figure 8.   Raman spectra at room temperature of As–S alloys containing 35–50 atomic % As

method of sample preparation. Such considerations are important in photoconductive and infrared device applications of this material.

The behavior of the As–Se system is similar but not identical. The concentration dependences of the melting point and the glass transition determined by Myers and Felty (7) are shown in Figure 9. The liquidus anomaly at 20 atomic % arsenic and the nonlinear form of the T_g curve is observed. In contrast to the As–S case, there is no compound formation at $As_{0.50}Se_{0.50}$ corresponding to As_4Se_4. These features are depicted in the Raman spectra shown in the central portion of Figure 10. (The other spectra shown in this figure refer to Se–S and Se–Te alloys, and these are considered later.) As the arsenic concentration increases, the shoulder at 239 cm^{-1} in the spectrum of vitreous selenium broadens, increases in intensity relative to the Se_8 peak at 250 cm^{-1}, and shifts to lower energy.

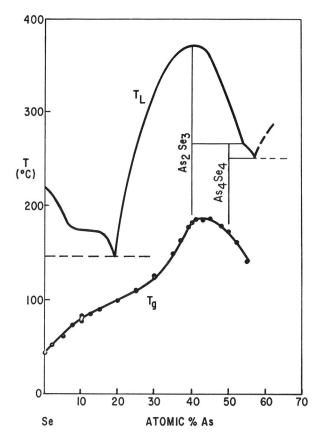

Figure 9.   Glass transition temperatures and liquidus temperatures for the As–Se system (7)

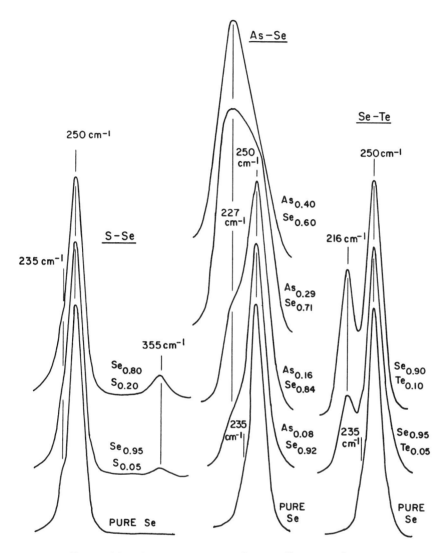

Figure 10. Raman spectra of binary alloys of selenium

Between 16 and 29 atomic % arsenic, the Se_8 rings disappear leaving only the broad spectral feature at 227 cm^{-1} associated with arsenic–selenium network polymer. This is the only feature in the spectrum of $As_{0.40}Se_{0.60}$. There are no other lines which are attributed to monomer species such as As_4Se_4.

A different situation exists in the Se–S and Se–Te alloys also shown in Figure 10. Here a new composition dependent spectral feature appears at 355 cm^{-1} for Se–S and at 216 cm^{-1} for Se–Te alloys. By mass-dependent

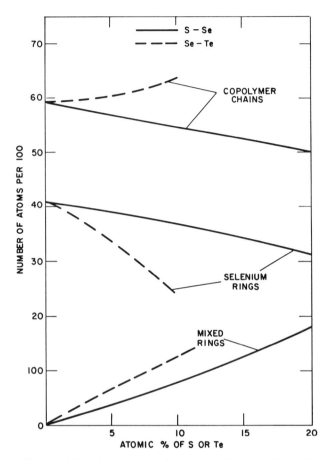

Figure 11. Atomic distributions in Se–S and Se–Te alloys

frequency correlations, these spectral features are assigned to mixed octa-atomic rings $Se_{8-x}S_x$ and $Se_{8-y}Te_y$, respectively, where x is probably 5 and y is probably 2. More complete evidence that mixed rings exist is seen by going to higher sulfur concentrations in the Se–S system. The exact values of x and y are not material.

The Tobolsky–Eisenberg polymerization theory was considered potentially useful for predicting molecular structure in vitreous materials. In principle, the copolymerization form of the theory described by Tobolsky and Owen (3) provides a way to calculate the concentrations of the various species present in equilibrium in the liquid phase. As a first approximation, the molecular constitution of the liquid phase at the M.P. is the structure retained in the vitreous phase upon rapid quenching to room temperature. A slight modification of the Tobolsky–Owen theory is required for Se–S and Se–Te alloys, because the copolymerization must involve pure Se_8 rings and mixed $Se_{8-x}S_x$ or $Se_{8-y}Te_y$ rings rather than Se_8 and S_8 or Te_8 as is required by strict adherence to the theory. In selecting the thermodynamic parameters appropriate to the mixed rings, the configurational change occurring during the polymerization

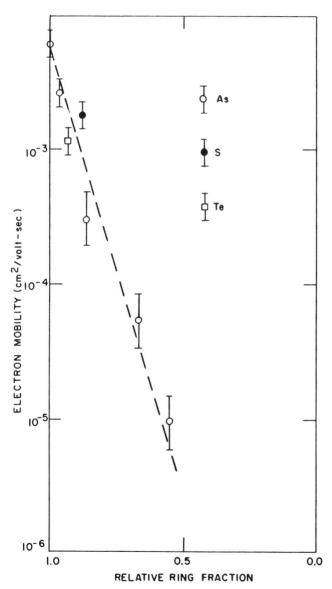

Figure 12.   Dependence of the electron drift mobility on the Se_8 ring concentration in binary alloys of Se

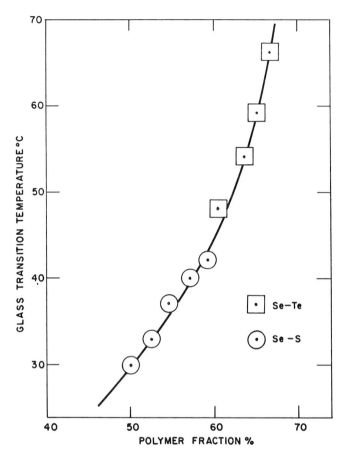

Journal of Physical Chemistry

*Figure 13.   Variation of the glass transition temperature with
polymer fraction in Se–S and Se–Te alloys* (9)

propagation reaction is assumed to be insensitive to the minority atoms
while the enthalpy change is assumed to be determined by the strength
of the weakest bond. The molecular constitution of Se–S and Se–Te
alloys is calculated now as a function of composition. The results are
shown in Figure 11. Note that the relative fractions of mixed octa-atomic
rings and pure Se_8 rings predicted theoretically agree qualitatively with
the relative intensities determined experimentally as spectral signatures
in Figure 10. The validity of the structure property correlations shown in
Figures 12 and 13 depends upon this point.

Figure 12 demonstrates a relationship between molecular structure
as represented by the Se_8 ring concentration and a physical property, the
electron drift mobility. The linear correlation confirms an earlier sup-

position by Schottmiller *et al.* (9), based on a comparison of the electrical properties of amorphous selenium and monoclinic selenium, that the electron mobility in selenium-based materials involves the Se_8 rings.

Figure 13 demonstrates a relationship between molecular structure, represented by the fraction of atoms present in polymeric form in Se–S and Se–Te alloys, and the glass transition temperature determined by Myers (10). The trend reflects the correlation made for binary alloys containing arsenic—*viz.*, the greater the fraction of polymer the higher

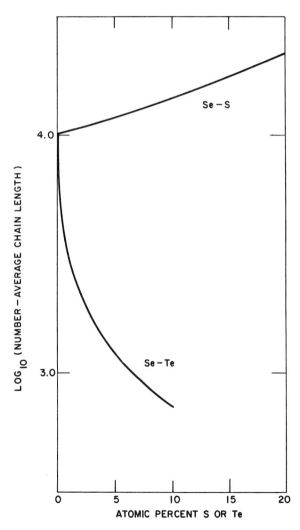

Figure 14. Variation of the number-average chain length with composition in Se–S and Se–Te alloys

the glass transition. This trend cannot hold for all alloy compositions, however, because of the competing effect of polymer chain length.

Figure 14 shows the composition–dependence of the number average chain length also calculated according to the modified Tobolsky–Owen theory. For a system such as Se–Te in which the polymer fraction is increasing linearly while the chain length is decreasing logarithmically, eventually a composition is reached where the influence of chain length dominates and T_g decreases.

In summary, laser Raman spectroscopy is useful for investigating the molecular microstructure of amorphous materials especially when reinforced by theoretical calculations. Correlations between spectroscopic data and the results of other physicochemical measurements are established. A detailed evaluation of the implications of these correlations must, however, be omitted to avoid intrusion upon areas of proprietary interest.

Literature Cited

(1) Tobolsky, A. V., Eisenberg, A., *J. Amer. Chem. Soc.* (1959) **31**, 780.
(2) Tobolsky, A. V., Eisenberg, A., *J. Polymer Sci.* (1960) **46**, 19.
(3) Tobolsky, A. V., Owen, A. D. T., *J. Polymer Sci.* (1962) **59**, 329.
(4) Carlson, C. F., U. S. Patent **2,297,691** (Oct. 6, 1942).
(5) Stuke, J., "Selenium," W. C. Cooper and R. E. Zingaro, Eds., Van Nostrand—Reinhold, New York, 1971.
(6) Hartke, G., Regensburger, P., *Phys. Rev.* (1965) **139**, A970.
(7) Myers, M. B., Felty, E. J., *Mater. Res. Bull.* (1967) **2**, 535.
(8) Ward, A. T., Myers, M. B., *J. Phys. Chem.* (1969) **73**, 1374.
(9) Ward, A. T., *J. Phys. Chem.* (1970) **74**, 4110.
(10) Schottmiller, J., Tabak, M., Lucovsky, G., Ward, A., *J. Non-Cryst. Solids* (1970) **4**, 80.
(11) Myers, M. B., Xerox Corp., private communication, 1970.

RECEIVED March 5, 1971.

Fluorinated Sulfide Polymers

C. G. KRESPAN, W. R. BRASEN, and H. N. CRIPPS

E. I. du Pont de Nemours and Co., Wilmington, Del. 19898

Tetrafluoroethylene and sulfur react at 445°C and 1 atm in a flow system to form tetrafluoro-1,2,3,4-tetrathiane and tetrafluoro-1,2,3-trithiolane in 60 and 10% yields, respectively, along with smaller amounts of tetrafluorothiirane and thiocarbonyl fluoride. Tetrafluorothiirane is prepared better from reaction of thiocarbonyl fluoride with hexafluoropropylene epoxide and thiocarbonyl fluoride from cracking of tetrafluoro-1,3-dithietane. The cyclic polysulfides are converted to unstable polymers of moderately high molecular weight and regular structure by weakly basic catalysts. Tetrafluorothiirane undergoes radical-catalyzed ring-opening polymerization to a stable monosulfide thermoplastic. Thiocarbonyl fluoride is converted by anionic or free radical initiators to a high molecular weight, moderately stable elastomer. Alternative routes to some of these polymers from bis(sulfenyl chlorides) provided fluorinated polysulfides, but of relatively low molecular weight.

The problem of synthesizing difluoromethylene chains joined by sulfur atoms, particularly in the medium to high molecular weight range, has been attacked most successfully by preparing individual monomers from tetrafluoroethylene and sulfur. The synthesis of several fluorocarbon sulfides and the techniques for polymerizing some of them are discussed below.

Molecular sulfur consists of stable eight-membered rings at low temperatures, but it is known to undergo homolytic cleavage in the melt with generation of free-radical terminated sulfur chains (*1*). In their reactivity toward tetrafluoroethylene, the radicals present in sulfur at 150°C or higher resemble thiyl radicals such as $CH_3S\cdot$ and $CF_3S\cdot$ (*2*), so that thermally initiated addition of sulfur to tetrafluoroethylene is accomplished easily. Under static conditions the reaction is carried out best in a suitable solvent such as carbon disulfide under pressure. Products formed depend critically on the ratio of sulfur to tetrafluoroethylene and, to a lesser extent, on temperature. The first intermediate is of the type 1, a diradical derived from addition of a sulfur chain to tetrafluoroethylene (3).

$$S_8 \rightleftharpoons \cdot S(S_x)S \cdot \quad \overset{CF_2=CF_2}{\longrightarrow} \quad \cdot S(S_x)SCF_2CF_2 \cdot$$

1

The presence of excess tetrafluoroethylene promotes reaction of the fluorocarbon radical end of 1 with another molecule of tetrafluoroethylene to form a new diradical, 2. Stabilization of 2 occurs by intramolecular attack of the carbon radical end on sulfur with formation of a favored thiolane ring, 3, and displacement of the sulfur chain. Yields of 20% of 3 have been attained.

$$1 + CF_2=CF_2 \rightarrow \cdot S(S_x)SCF_2CF_2CF_2CF_2 \cdot \rightarrow$$

2   3

Evidence against an ionic mechanism in such reactions is the observation that at 120°C, where reaction is slow, no acceleration was noted with added triethylamine. Evidence for a free radical mechanism is found in response to the presence of an olefin as a third reactant. Hydrocarbon olefins with one substituent intercept intermediate 1 more efficiently than does tetrafluoroethylene, and they give partially fluorinated thiolanes of type 4. Regardless of the electronegativity of the olefinic substituent, yields are good (40–60%), and products have the substituent on a carbon adjacent to sulfur in the ring (4). An intermediate of the type 5, in which nearly any substituent provides some stabilization to the adjacent radical center, is implicated. Cationic or anionic intermediates result in pronounced deviations from the results observed.

$$1 + CH_2=CHX \rightarrow \cdot S(S_x)SCF_2CF_2CH_2\overset{\cdot}{C}HX \rightarrow$$

5   4

$(X = CH_3, Cl, CN, OCOCH_3, etc.)$

Returning to the two-component reaction at 150°C, excess sulfur promotes formation of cyclic sulfides and polymers rich in sulfur; a major product is the cyclic tetrathiane 6 ($n = 4$). As the temperature of such long term static reactions is raised to 300°C, product composition changes to a predominance of the more stable cyclic sulfides, **7** and **8**. Octafluoro-dithiane, **7**, is obtained easily under these conditions in 50–60% yield. Pyrolyses of preformed cyclic polysulfides such as **6** ($n = 4$) and of polymeric polysulfides carried out at 300°C similarly result in formation of **7** and **8** along with free sulfur.

Pyrolytic conversion of **6** to **7** with extrusion of sulfur is a slow reaction which apparently involves in the latter stages an unusual attack of thiyl radical on carbon to form the monosulfide structure. Pyrolysis of **8** to **7**, also demonstrated separately, is a more direct example of conversion of a fluoroalkyl disulfide to monosulfide by intramolecular attack of thiyl radical on carbon. The dithiane **7**, once formed, is stable at 300°C, indicating that the CF_2–S bond in these compounds is not subject to homolysis. The related formation of an acyclic monosulfide—e.g., CF_3SCF_3—from an acyclic disulfide—e.g., CF_3SSCF_3—requires photochemical activation or temperatures higher than 300°C.

The unstable cyclic polysulfides are prepared best in a flow system by passing gaseous tetrafluoroethylene through vapors of refluxing sulfur (445°C) and continuously withdrawing product (5). This technique allows isolation of the tetrathiane **6** ($n = 4$) in 60% yield as a distillable pale yellow oil which is fairly stable in the absence of base or light. A lesser product is the cyclic trithiolane **6** ($n = 3$), a yellow oil which is distilled in small amounts but polymerizes with such ease that it cannot be stored. Minor products are thiocarbonylfluoride, **9**, formed by rupture of the carbon–carbon bond of **6** at the high temperatures involved, and tetrafluorothiirane, **6** ($n = 1$).

In the vapor phase 1 tends to undergo intramolecular attack of the fluorocarbon radical end on various sulfur atoms along the chain. Preference for the six-membered ring is strong, making this the best route to 6 ($n = 4$). Attack at the third sulfur atom to give 6 ($n = 3$) is favored less (10% yield), presumably because the more nearly coplanar five-membered ring forces the ring atoms into smaller dihedral angles, resulting in destabilizing repulsions between unshared electron pairs on adjacent sulfur atoms. The observed difference in ease of opening the two rings fits this picture, as do the ^{19}F NMR spectra. The tetrathiane exists in a stable puckered conformation, having distinguishable fluorine atoms, while the trithiolane showed a broad singlet indicating a more planar structure. Continuing the trend of polysulfide instability with smaller ring size, the cyclic disulfide 6 ($n = 2$) proved too unstable to isolate under normal conditions. However, it is capable of short-lived existence and probably is formed in low yield from excess sulfur vapor and tetrafluoroethylene.

Tetrafluorothiirane, 6 ($n = 1$), lacks a sulfur–sulfur bond and is fairly stable when formed, but it is formed in only 1–2% yield in the above reaction. A more useful synthesis of this monomer is the reaction of thiocarbonyl fluoride with hexafluoropropylene epoxide at 175°C. At this temperature hexafluoropropylene epoxide serves as a source of difluoromethylene, which adds to the double bond of thiocarbonyl fluoride to give the thiirane. If close control of conditions is maintained, yields are close to 40% (6).

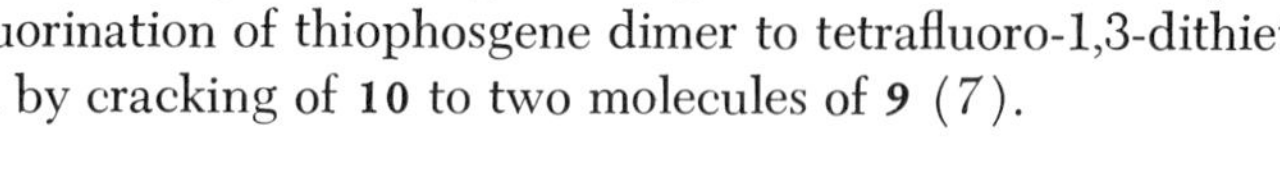

$$F_2C{=}S + F_2 \underset{O}{\diagdown\diagup} \quad FCF_3 \rightarrow \underline{6}\ (n = 1) + CF_3COF$$

Thiocarbonyl fluoride, 9, is prepared in good yield by pyrolysis of 6 at ca. 600°C in the vapor phase or by carrying out the reaction of sulfur and tetrafluoroethylene at higher temperatures. A cleaner synthesis involves fluorination of thiophosgene dimer to tetrafluoro-1,3-dithietane, 10, followed by cracking of 10 to two molecules of 9 (7).

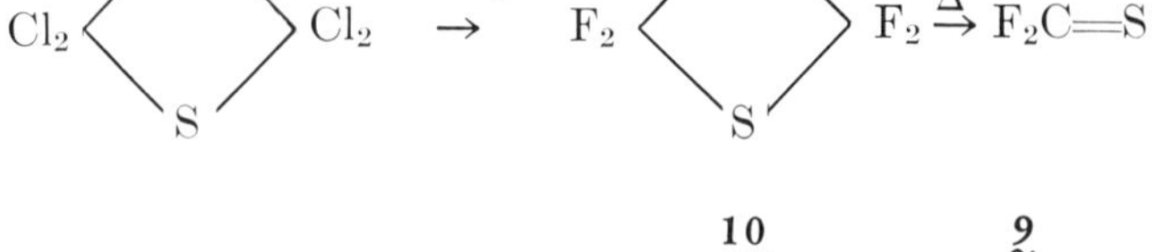

$$\underset{\underline{10}}{\qquad} \qquad \underset{\underline{9}}{\qquad}$$

Of the products available from this chemistry, five polymerize under appropriate conditions. The polysulfides, 6 ($n = 4$), 6 ($n = 3$), and 8,

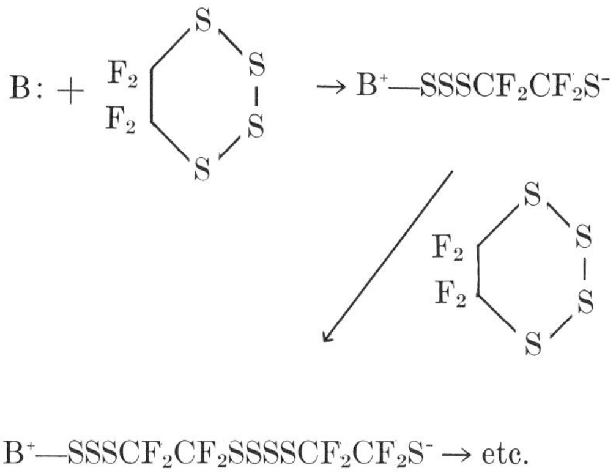

$$B: + \begin{matrix}F_2 \\ F_2\end{matrix} \bigg[\text{(ring)} \bigg] \rightarrow B^+\text{---}SSSCF_2CF_2S^-$$

$$B^+\text{---}SSSCF_2CF_2SSSSCF_2CF_2S^- \rightarrow \text{etc.}$$

$$\underset{\sim\sim}{\textbf{11}}$$

are polymerized most effectively by weak bases and yield highly crystalline polymers indicating a regular structure. The formation of a polytetrasulfide, **11**, rather than mixed polysulfides from **6** ($n = 4$), for example, is understood by considering the most likely point of attack in the propagation step. The anionic terminal sulfur of the growing chain attacks a ring sulfur atom one removed from the fluoroalkyl moiety to generate the most stable mercaptide ion—*i.e.*, a fluoroalkyl mercaptide.

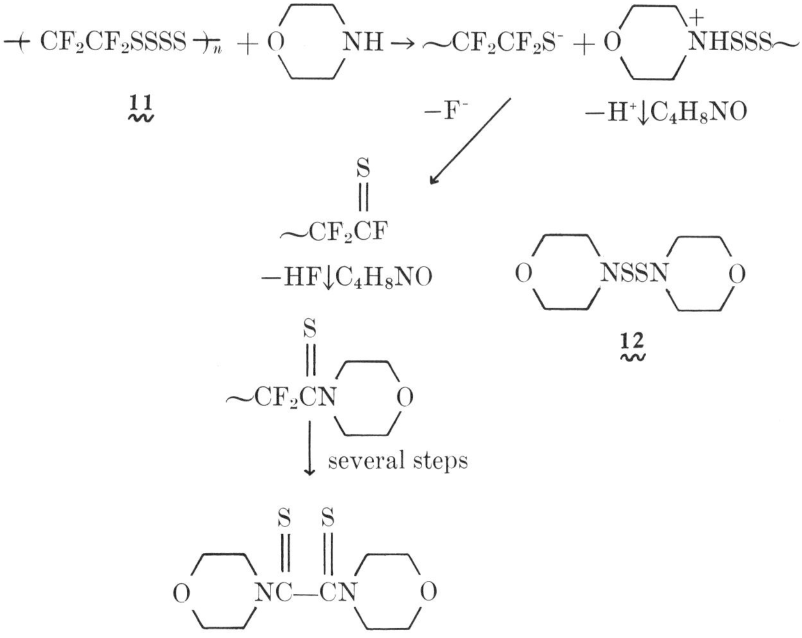

$$\text{---}\hspace{-2pt}\left(CF_2CF_2SSSS\right)_n + O\!\!\left(\text{ring}\right)\!\!NH \rightarrow \sim\!\!CF_2CF_2S^- + O\!\!\left(\text{ring}\right)\!\!\overset{+}{N}HSSS\!\!\sim$$

$$\underset{\sim\sim}{\textbf{11}} \hspace{3cm} -F^- \hspace{2cm} -H^+\!\downarrow C_4H_8NO$$

$$\overset{S}{\underset{\parallel}{\sim\!\!CF_2CF}} \hspace{3cm} O\!\!\left(\text{ring}\right)\!\!NSSN\!\!\left(\text{ring}\right)\!\!O$$

$$-HF\!\downarrow C_4H_8NO \hspace{4cm} \underset{\sim\sim}{\textbf{12}}$$

$$\overset{S}{\underset{\parallel}{\sim\!\!CF_2CN}}\!\!\left(\text{ring}\right)\!\!O$$

$$\downarrow \text{several steps}$$

$$O\!\!\left(\text{ring}\right)\!\!\overset{S\ \ S}{\underset{}{NC\!\!\overset{\parallel\ \ \parallel}{\text{---}}CN}}\!\!\left(\text{ring}\right)\!\!O$$

$$\underset{\sim\sim}{\textbf{13}}$$

Thus, the result of adding 6 ($n = 4$) slowly to stirred acetonitrile maintained at $-40\,^\circ$C is an 85% yield of tough, white polymer, crystalline melting point 56°C, $\eta_{inh} = 1.0$ at 25°C in toluene, and a singlet for the ^{19}F nmr spectrum. X-ray studies indicated a highly crystalline structure. Degradation of the polymer with morpholine confirmed the poly(tetrasulfide) structure in that good yields of dimorpholine disulfide, 12, and the dithiooxamide 13 are obtained; the reaction involves the same type of attack on the polymer to displace fluoroalkyl mercaptide as does the ring opening polymerization step.

Conditions suitable for polymerizating the tetrathiane are those which minimize loss of fluoride ion from the fluoroalkyl mercaptide intermediate, catalysis by weak bases such as acetonitrile, acetone, or ethanol at low temperatures. Similar systems are effective in polymerizing trithiolane 6 ($n = 3$), to a white polymer, 14, crystalline melting point 95–100°C, $\eta_{inh} = 1.2$ in toluene at 25°C. The trithiepane 8, however, was polymerized best by a stronger base, triethylamine, at low temperature, but the insoluble product, 15, pressed into weak films of relatively low molecular weight.

$$(\; CF_2CF_2SSS \;)_n \qquad\qquad (\; CF_2CF_2SCF_2CF_2SS \;)_n$$

$$14 \qquad\qquad\qquad\qquad 15$$

The three polymers under discussion, 11, 14, and 15, are all subject to attack by base, rupture occurring at sulfur–sulfur bonds with loss of fluoride ion. Moreover, the sulfur–sulfur bonds, particularly in 11 and 14, are susceptible to thermally and photochemically induced homolysis. Samples of high η_{inh} lose viscosity on standing, most rapidly when exposed to light, and lability of the sulfur–sulfur bonds at 25°C is illustrated by a slow formation of sulfur crystals on pressed films. At temperatures above 200°C, rapid pyrolysis of 11 occurs to give regenerated monomer 6 ($n = 4$) as the main product. From the general thermal stability of the sulfur–sulfur bond in fluoroalkyl disulfides, the bond ruptured homolytically in 11 is expected to be preferentially that which gives $\cdot SSCF_2\sim$ radicals. This appears to be the case with the monomer, tetrathiane 6 ($n=4$) since its reaction with tetrafluoroethylene at 200°C gives directly a high yield of the regularly repeating poly(disulfide), 16.

$$+ \; CF_2{=}CF_2 \rightarrow +CF_2CF_2SS +_n$$

$$\underset{\sim\!\sim}{16}$$

Formation of **16** may proceed through eight-membered and four-membered ring intermediates, species which are apparently unstable with respect to the polymer. A demonstration of this ready polymerizability is volatilization of **16** at 250°C under reduced pressure resulting in volatiles indicated by mass spectrometry to be **6** ($n = 2$) and **17**. Collection of the volatiles in a cold trap allows complete transfer of polymer **16** to the trap with no sign of low molecular weight species in the condensate. Trapping of the diethietane **6** ($n = 2$) is possible; pyrolysis of **16** at 300°C in the presence of hexafluoropropylene resulted in the adduct **18**.

These reactions suggest that the major routes to the dithiane **7** and trithiepane **8** are by additions of tetrafluoroethylene to the 1,2-dithietane **6** ($n = 2$) and the trithiolane **6** ($n = 3$), respectively.

Polymerization of the cyclic polysulfides is carried out also by using free radical catalysis but with less satisfactory results than with basic catalysts. Tetrafluorothiirane, 6 ($n = 1$), on the other hand, is polymerized to waxy solids with fluoride ion systems at low temperatures but goes cleanly to high molecular weight polymer 19 with free radical catalysis. Initiating with *trans*-dinitrogen difluoride at 60°C or bis(trifluoromethyl) disulfide and ultraviolet light gives a high yield of white, tough, highly ordered polymer, crystalline melting point around 175°C. Such indications of a regular monosulfide structure are confirmed by 19 resisting attack by base, showing no significant number of disulfide bonds present.

$$\text{F}_2 \diagdown \diagup \text{F}_2 \quad \xrightarrow{\text{R}\cdot} \quad [\text{RSCF}_2\text{CF}_2\cdot] \quad \xrightarrow{\quad \text{F}_2 \diagdown \diagup \text{F}_2 \quad}$$

$$\underset{\sim}{6} \ (n = 1)$$

$$\text{+ CF}_2\text{CF}_2\text{S +}_n$$

$$\underset{\sim\sim}{19}$$

The propagation step in this free radical polymerization has been demonstrated to involve radical attack on episulfide sulfur (as depicted above) by telomerization studies with cyclohexane. From the structures of the two major products, 20 and 21, cyclohexyl radical attacks at the sulfur atom by displacing fluoroalkyl radical. This new radical, a prototype of the growing polymer chain, then abstracts hydrogen from cyclohexane to form 20 or attacks another molecule of 6 ($n = 1$) at sulfur to give 21.

$$6 \ (n = 1) + \text{C}_6\text{H}_{12} \xrightarrow{\text{R}\cdot} \text{C}_6\text{H}_{11}\text{SCF}_2\text{CF}_2\text{H} + \text{C}_6\text{H}_{11}\text{SCF}_2\text{CF}_2\text{SCF}_2\text{CF}_2\text{H}$$

$$\qquad\qquad\qquad\qquad\qquad\qquad\quad 20 \qquad\qquad\qquad\qquad 21$$

Monosulfide polymer 19 is the most stable of the polymers discussed, not only in its resistance to base, oxidizing acids, and light but also in thermal stability. Degradation is slow at 300°C, but at 350°C chain scission results in an unzippering to evolve tetrafluoroethylene and dithietane 6 ($n = 2$). The latter gives rise to polydisulfide 16 and may cycloadd to tetrafluoroethylene to form dithiane 7, the third major product. More likely under these conditions, however, is direct formation of 7 by a backbiting mechanism as illustrated below.

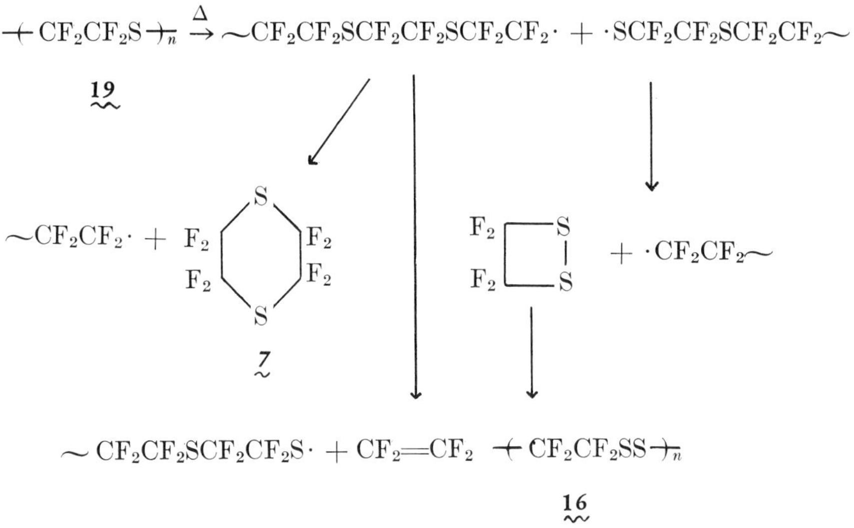

$$+CF_2CF_2S+_n \xrightarrow{\Delta} \sim CF_2CF_2SCF_2CF_2SCF_2CF_2\cdot + \cdot SCF_2CF_2SCF_2CF_2\sim$$

19

$$\sim CF_2CF_2\cdot + F_2 \underset{S}{\overset{S}{\diamondsuit}} F_2 \qquad F_2 \underset{S}{\overset{S}{\square}} F_2 + \cdot CF_2CF_2\sim$$

7

$$\sim CF_2CF_2SCF_2CF_2S\cdot + CF_2{=}CF_2 \quad +CF_2CF_2SS+_n$$

16

In accord with this interpretation is the formation of large amounts (25–30%) of octafluorothiolane, **3**, in copolymerizations of the thiirane with tetrafluoroethylene. After incorporation of a molecule of **6** ($n = 1$) into the growing chain, addition of tetrafluoroethylene creates a four-carbon segment with a radical end. Intramolecular attack by this radical on sulfur generates thiolane **3**.

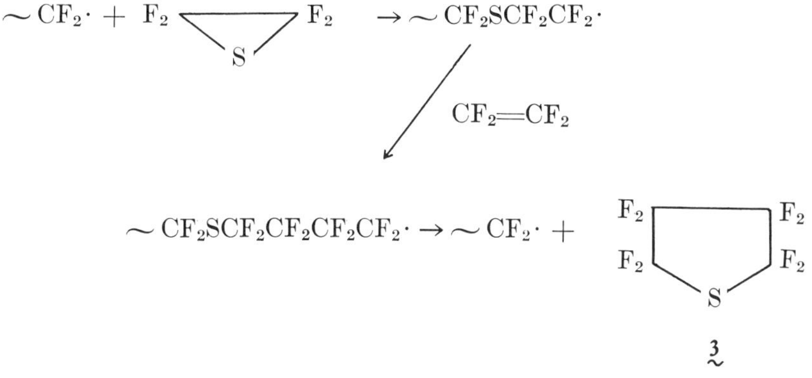

$$\sim CF_2\cdot + F_2 \overset{\triangle}{\underset{S}{}} F_2 \rightarrow \sim CF_2SCF_2CF_2\cdot$$

$$CF_2{=}CF_2$$

$$\sim CF_2SCF_2CF_2CF_2CF_2\cdot \rightarrow \sim CF_2\cdot + F_2 \underset{S}{\overset{\square}{}} F_2$$

3

Copolymerization of **6** ($n = 1$) with hydrocarbon olefins proceeds readily, so that complications of the sort encountered with tetrafluoroethylene are suppressed. For example, copolymerization with propylene has given a 1:1 elastomeric product of high molecular weight.

Thiocarbonyl fluoride, **9**, can be polymerized with either anionic or radical catalysts. Low temperatures are required to attain high molecular weights. Weak bases such as dimethylformamide at $-50°$ to $-80°C$ give polymer 22 with a crystalline–amorphous transition of $35°C$ and $\eta_{inh} = 2.0–5.0$ in chloroform at $30°C$ (8). Essentially the same polymer is obtained at $-60°$ to $-80°C$ using trialkylboron–oxygen as radical initiator (9). With either type of initiation, higher temperatures result in reduced molecular weight.

$$CF_2{=}S \rightarrow (CF_2S)_n$$
$$9 \qquad\qquad 22$$

The general sulfur atoms in 22 render it less stable than 19, so that, while quite stable toward acids, amines degrade it rapidly. Thermal decomposition occurs above about $175°C$ by unzippering to regenerate monomer.

As with thiirane 6 ($n = 1$), copolymerization of 9 with vinyl monomers generally proceeds well under free radical conditions. A number of copolymers have been prepared.

A different approach to long chain fluorinated sulfides is through the reactions of fluorinated sulfenyl chlorides. Such sulfenyl chlorides are available from direct chlorination of the cyclic polysulfides described above (10). Photochemical chlorination of trithiepane 8 serves to produce bis(sulfenyl) chloride 23. Both 6 ($n = 4$) and 6 ($n = 3$) give bis(sulfenyl) chloride 24 with chlorine in an inert medium without need for photoactivation. Incomplete chlorination leads to disulfur chlorides as by-products in the latter cases.

$$ClSCF_2CF_2SCF_2CF_2SCl \qquad\qquad ClSCF_2CF_2SCl$$
$$23 \qquad\qquad\qquad\qquad 24$$

Condensation polymerization of the bis(sulfenyl) chlorides is effected by treatment with mercury or potassium iodide. However, the polymers are of lower molecular weight than the analogous products obtained directly from the cyclic polysulfides.

$$\overset{Hg}{}$$
$$ClSCF_2CF_2SCl \rightarrow (CF_2CF_2SS)_n$$
$$24 \qquad\qquad\qquad 16$$

Addition of the fluorinated sulfenyl halides to olefinic bonds occurs readily; for example, with ethylene and 24 the reaction gives 25. Similar reactions occur with unsaturated polymers to generate crosslinks.

$$24 + CH_2{=}CH_2 \rightarrow ClCH_2CH_2SCF_2CF_2SCH_2CH_2Cl$$

25

Chlorination of a cyclic polysulfide in the presence of water results in further oxidizing of the sulfenyl chloride groups to sulfonyl chlorides, for instance 26 prepared from 6 ($n = 4$).

$$ClSO_2CF_2CF_2SO_2Cl$$

26

Use of this sulfonyl chloride as an intermediate to sulfonamides is hindered by a tendency to undergo fragmentation. Reaction of 26 with aniline gives tetrafluoroethylene as a major product.

Summary

The presence of isolated sulfur atoms in a long difluoromethylene chain confers increased tractability to the polymer at the expense of thermal stability. Long term exposure to temperatures near 300°C results in slow degradation, apparently initiated by scission of C–S bonds in the polymer backbone. Although the thermal stability of such fluorinated sulfide polymers is closer to that of many well-known polymethylenes and polyamides, resistance to chemical attack remains high. Larger amounts of sulfur incorporated into the polymer backbone, particularly in the form of disulfide and higher polysulfide units, further decreases thermal stability and gives rise to marked susceptibility to base attack.

Literature Cited

(1) Schmidt, M., *Angew. Chem.* (1961) **73**, 394.
(2) Harris, Jr., J. F., Stacey, F. W., *J. Amer. Chem. Soc.* (1961) **83**, 840.
(3) Krespan, C. G., Langkammerer, C. M., *J. Org. Chem.* (1962) **27**, 3584.
(4) Krespan, C. G., *J. Org. Chem.* (1962) **27**, 3588.
(5) Krespan, C. G., Brasen, W. R., *J. Org. Chem.* (1962) **27**, 3995.
(6) Brasen, W. R., Cripps, H. N., Bottomley, C. G., Farlow, M. W., Krespan, C. G., *J. Org. Chem.* (1965) **30**, 4188.
(7) Middleton, W. J., Howard, E. G., Sharkey, W. H., *J. Org. Chem.* (1965) **30**, 1375.
(8) Middleton, W. J., Jacobson, H. W., Putnam, R. E., Walter, H. C., Pye, D. G., Sharkey, W. H., *J. Polymer Sci. Part A* (1965) **3**, 4115.
(9) Barney, A. L., Bruce, J. M., Coker, J. M., Jacobson, H. W., Sharkey, W. H., *J. Polymer Sci. Part A-1* (1966) **4**, 2617.
(10) Krespan, C. G., U. S. Patent **3,099,688** (1963).

RECEIVED March 5, 1971

14

Electrical Conductivity of Liquid Sulfur and Sulfur–Phosphorus Mixtures

R. K. STEUNENBERG, C. TRAPP,[1] R. M. YONCO, and E. J. CAIRNS

Chemical Engineering Division, Argonne National Laboratory, 9700 South Cass Ave., Argonne, Ill. 60439

An investigation has been conducted on the electrical conductivities of liquid sulfur and sulfur–phosphorus mixtures. Both ac and dc measurements were made, using bridge techniques. The dc measurements gave conductivity values of 2.0×10^{-10} and 3.5×10^{-9} ohm^{-1} cm^{-1} for sulfur and P_4S_3, respectively, at 350°C, with corresponding activation energies of 37.2 kcal/gram-atom and 7.4 kcal/mole. The conductivity of sulfur appeared to decrease with increasing ac frequency, but P_4S_3 did not exhibit this effect. Small additions of phosphorus in the form of P_4S_{10} caused an increase in sulfur's conductivity.

An investigation has been conducted on the electrical conductivities of liquid sulfur and some sulfur–phosphorus mixtures. The purpose of this work was: (1) to obtain additional data on the conductivities of liquid semiconductors, which are of theoretical interest, and (2) to investigate the possibility of using these materials as the cathode reactant in lithium–chalcogen electrochemical cells. It is difficult to obtain reliable conductivity data on these systems because of their low conductivities, the strong influence of impurities on their conductivity, and corrosion problems associated with most electrode materials.

The conductivity of liquid sulfur has been measured by Feher and Lutz (1), Gordon (2), and Watanabe and Tamaki (3). Also Poulis and Massen (4) have calculated conductivity values for sulfur at 231° and 360°C from the dielectric-constant data of Curtis (5). The conductivities reported by Feher and Lutz are much lower than those of Watanabe and Tamaki, but the former data were obtained at lower temperatures than

---

[1] Present address: University of Louisville, Louisville, Ky. 40208.

the latter. A linear extrapolation of results obtained by Kraus and Johnson (6) on the tellurium–sulfur system to pure sulfur agrees with an extrapolation of the Feher and Lutz data to higher temperatures. Only Watanabe and Tamaki had made measurements above 300°C, and their results disagree with the others. The present investigations on sulfur were conducted in the higher temperature region.

No information has been found in the literature on the electrical conductivities of sulfur–phosphorus mixtures. The following studies do not cover the entire sulfur–phosphorus system; they are limited to measurements on sulfur, P_4S_3, and two intermediate compositions in the sulfur-rich region of the system.

Experimental

Materials. Sulfur was obtained from the American Smelting and Refining Co. with a purity specification of 99.999+%. No further purification was attempted. The P_4S_3 was procured from the Eastman Kodak Co. One batch of this material, which was purified by extraction with CS_2 followed by sublimation, had a melting point of 171.4 ± 0.1°C. The best literature value for the melting point of P_4S_3 appears to be 174 ± 1°C (7). A second batch was treated with aqueous $NaHCO_3$ solution, washed with water, vacuum dried, and then extracted with CS_2 and recrystallized under an argon atmosphere. The melting point of this product was 173.8 ± 0.1°C, which suggests that it was of higher purity than the previous batch. The P_4S_{10} used to prepare the sulfur-rich mixtures of sulfur and phosphorus, which was obtained from the Eastman Kodak Co., was purified by CS_2 extraction and sublimation. The sulfur used in these mixtures was Mallinckrodt sublimed N. F. grade material.

Conductivity Cells. The measurements on sulfur were made with a quartz cell with tungsten leads sealed through the bottom. The electrodes consisted of spectographic graphite cylinders 6.3 mm in diameter and 2.5 cm high with an interelectrode distance of about 1 cm between the axes, giving a cell constant of 0.124 cm^{-1}. The cell used in the P_4S_3 measurements was made also of quartz with tungsten leads passing through the bottom. Within the cell the leads were protected with borosilicate glass sheaths. The electrons were vitreous carbon disks 2.54 cm in diameter and 3.2 mm thick, which were obtained from the Beckwith Carbon Corp. The interelectrode distance between the faces of the disks was ~0.8 mm, with a resulting cell constant of 0.0215 cm^{-1}. The measurements on the sulfur-rich sulfur–phosphorus mixtures, which were made earlier, utilized niobium electrodes 1 cm square with an interelectrode distance of about 0.5 mm to give a cell constant of 0.0531 cm^{-1}. The electrodes were attached to niobium leads and were inserted from above into a borosilicate glass cell.

Electrical Measurements. Both ac and dc conductivity measurements were made on elemental sulfur and P_4S_3. The ac measurements were made with an Electro Scientific Industries model 290A impedance bridge in conjunction with a model 860A generator/detector at frequencies of 0.1, 1.0, and 10.0 kHz. A compensating variable capacitor was used either in series or in parallel with the variable resistance arm of the bridge. The

SULFUR RESEARCH TRENDS

generator signal and the detector output from the bridge were applied to the X and Y inputs of an oscilloscope, which served as a null detector for both resistive and capacitive balance of the circuit. The dc measurements on sulfur and P_4S_3 were made with the same bridge, using 1.5-, 6-, or 18-volt dry cells as the power supply. A Leeds and Northrup model 9835B microvolt amplifier served as a null detector. The temperature was measured with a Pt/Pt–10% Rh thermocouple, which led to a Leeds and Northrup type K-3 potentiometer, a Leeds and Northrup model 9835B microvolt amplifier, and a strip-chart recorder. The precision of the temperature measurements with this arrangement was better than 0.01°C. The thermocouple was calibrated with National Bureau of Standards tin (231.88°C), lead (327.4°C), and zinc (419.50°C) to give an accuracy of ±0.1°C in the temperature measurements.

The earlier measurements on the sulfur-rich portion of the sulfur–phosphorus system were made with ac at a frequency of 1.0 kHz, using a Leeds and Northrup Jones bridge (No. 4666) in conjunction with an oscillator (No. 9842) and amplifier (No. 9847). The temperature was measured with an uncalibrated chromel/alumel thermocouple with a precision of 0.1°C and an estimated accuracy of ±5°C.

Procedure. The conductivity cells were calibrated with standard aqueous KCl solutions by the usual method (8). The conductivity of the deionized water used to prepare the KCl solutions was measured, and the appropriate correction was applied in the cell calibration. For the measurements on sulfur and P_4S_3, the cell was filled with sufficient solid material so that upon melting the liquid level was well above the electrodes. The required height of the liquid level above the electrodes to avoid affecting the conductivity measurements was established during the calibration with KCl solutions. The conductivity cell was operated in a furnace well attached to the floor of a high purity, helium atmosphere glovebox (9) to avoid possible atmospheric contamination. When the system had come to equilibrium at the desired temperature, the conductivity measurements were made with direct current and with alternating current at the various frequencies. The measurements were then repeated at different temperatures. In most cases, it was necessary to use a known shunt resistance (7.5×10^5 or 1.1×10^6 ohm) in parallel with the cell to decrease the total resistance to a value within the range of the bridge ($< 1.2 \times 10^6$ ohm). The compensating capacitance required to achieve a balance in the ac measurements was usually between 0 and 1200 pF, depending on the frequency.

The dc measurements on P_4S_3 were supplemented by a few additional determinations of the current as a function of time at various voltages. These measurements were made with a simple circuit in which the voltage across a known resistance in series with the cell was determined by a millivolt strip-chart recorder with a variable zero suppression.

In the case of the sulfur-rich mixtures of sulfur and phosphorus, approximately 120 grams of the mixture was placed in a borosilicate glass beaker, which was covered with a niobium plate at the higher temperatures (>350°C) to decrease sulfur losses by vaporization. The cell was placed in a 500 ml alumina secondary container, and this assembly was located inside a sealed stainless steel vessel which was heated by Nichrome resistance wire. The cover of the vessel had access ports for a stirring rod, thermocouple, and the electrode assembly, which was lowered into the melt. The vessel was evacuated and filled with argon prior to the measurements. The conductivity determinations were made with the Jones bridge at different temperatures and with a frequency of 1 kHz. A 5.8×10^5-ohm shunt resistance was used to maintain the measured resistance within the range of the bridge (6×10^5 ohm).

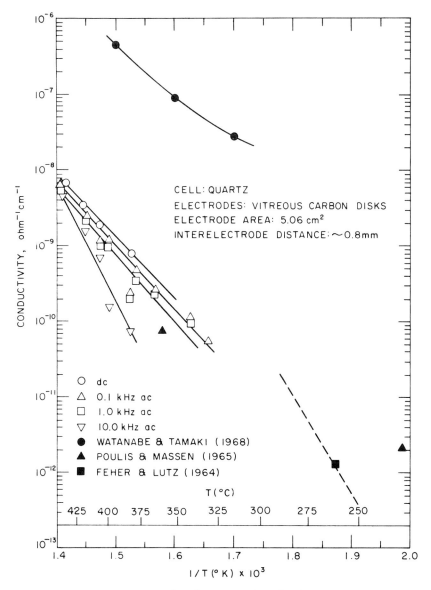

Figure 1. Conductivity of liquid sulfur

Results

Sulfur. The conductivity of sulfur was measured over the temperature range of 330°–435°C. The results are presented in Figure 1, along with the results of Watanabe and Tamaki (3), the two values reported by Poulis and Massen (4), and a value reported by Feher and Lutz (1). The dashed line through the Feher and Lutz value has a slope corresponding to their reported activation energy of 2.9 ev. The present data show the expected linear relationship between log κ and $1/T(°K)$, where κ represents the conductivity in ohm^{-1} cm^{-1}. Although there is considerable scatter in the data, the conductivity of sulfur appears to decrease with increasing frequency, and extrapolation of the ac results to zero frequency results in reasonable agreement with the dc data. The temperature dependence of the dc results corresponds to an activation energy of 37.2 kcal/gram-atom, or 1.62 ev. The activation energy, ΔE, was derived from the slope of the log κ vs. $1/T$ plot, which corresponds to $-\Delta E/2.303\ R$.

P_4S_3. Two sets of measurements were made with P_4S_3, one with the first batch of material (m.p. 171.4°C), and the other with the second, more highly purified batch (m.p. 173.8°C). The results, shown in Figure 2, indicate a linear dependence of log κ upon $1/T(°K)$, and the conductivity appears to be independent of ac frequency. Furthermore, the ac and dc results are identical within the experimental uncertainty. The temperature dependence of the conductivity for the higher purity batch of P_4S_3 corresponds to an activation energy of 7.4 kcal/mole or 0.32 ev.

Additional dc measurements were made on the higher purity P_4S_3 at 308°C to determine whether the dc conductivity is time dependent and whether a decomposition voltage could be observed. At an applied potential of 1.5 volts, the current decreased to about 94% of its initial value after approximately 10 min and then remained constant. When the potential was removed for 10 min and reapplied, the current started at about 97% of its original initial value and decreased again to about 94%. Reversing the polarity caused the current to return to the original initial value, followed again by a decrease to about 94% of that value.

Further measurements were made, determining the current over a period of time at various applied potentials ranging from 1–16 volts. At the lower voltages (below about 8 volts), the current behaved essentially as described above. At the higher voltages, however, the initial decrease in the current was followed by an increase to a level somewhat higher than the initial value. When the current leveled off at this higher value with applied potentials of 12 and 16 volts, it showed a noticeable, although not large, fluctuation. Figure 3 shows the observed relationship between the current density and the voltage in these experiments. The upper line

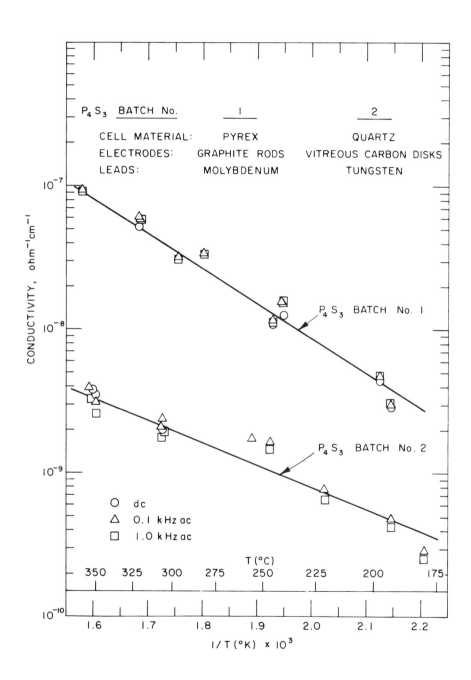

Figure 2. Conductivity of liquid P₄S₃

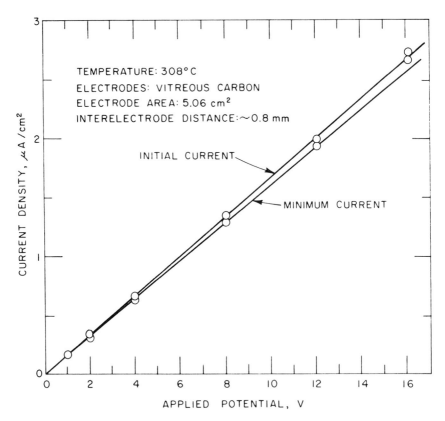

Figure 3.   Voltage-current density characteristics of liquid P$_4$S$_3$

represents the initial current, and the lower one the minimum current. The system appears to obey Ohm's Law up to perhaps 10 volts and then begins to deviate slightly in the direction of higher conductivity. However, there is no clear indication of a decomposition potential.

Phosphorus–Sulfur Mixtures. Figure 4 illustrates the effect of phosphorus concentration on the conductivity of the sulfur–phosphorus system in the sulfur-rich region. These data were obtained in earlier experiments prior to the present conductivity measurements on sulfur. The sulfur used in the earlier measurements was of somewhat lower purity and higher conductivity. The data in Figure 4 were corrected for the difference in conductivities of the two types of sulfur, assuming that the conductivities are additive. The conductivity shows a sharp rise with increasing phosphorus content up to a concentration of about 9 atom %, which was the maximum phosphorus concentration investigated in this set of measurements. The activation energies for the sulfur–phosphorus mixtures containing 2.2 and 9.3 atom % phosphorus are 17.2 and 15.4 kcal/mole (0.75

and 0.65 ev), respectively. These data are not regarded as being highly reliable or accurate, but they show a general trend that is believed to be valid.

Discussion and Conclusions

Within the temperature range of the measurements, the conductivity data on sulfur show a linear relationship between log κ and $1/T$ (°K).

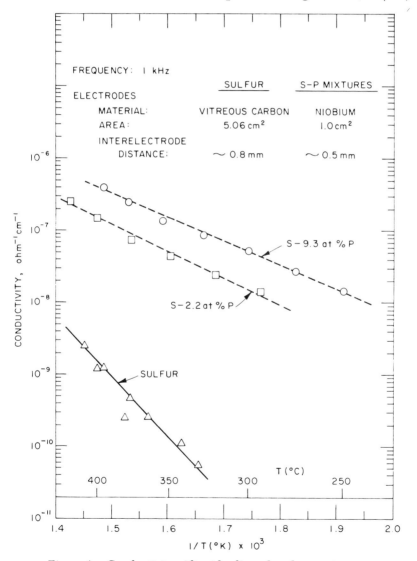

Figure 4.   Conductivity of liquid sulfur–phosphorus mixtures

The conductivity increases with decreasing frequency, and as the frequency decreases to zero, the ac results agree well with the dc data. It may be questionable whether the observed frequency dependence is truly a characteristic of the sulfur conductivity or if it could result from the experimental technique and equipment used in the measurements. However, the fact that a frequency dependence was not evident in the measurements on P_4S_3 under similar conditions suggests that the effect is real.

The dc conductivity data for sulfur extrapolate well to the lower temperature value of 1.35×10^{-12} ohm^{-1} cm^{-1} at 260°C reported by Feher and Lutz (1), but the activation energy of 1.62 ev (37.2 kcal/gram-atom) is considerably lower than their value of 2.9 ev. The results of Watanabe and Tamaki (3) in Figure 1 were read from their plot of resistivity vs. composition of selenium–sulfur mixtures at three temperatures, which they extrapolated to pure sulfur. The present values of the conductivity of sulfur are much lower than the data of Watanabe and Tamaki. Although their data show curvature in the log κ vs. $1/T$ (°K) plot in Figure 1, the activation energy corresponding to the slope of the higher temperature data in this plot is in reasonable agreement with that of the present data. Poulis and Massen (4) have calculated the conductivity of sulfur at a frequency of 312 Hz for two temperatures, using the dielectric constant and power factor results of Curtis (5). These results fall in the same general region as the present data but show considerably lesss dependence of the conductivity upon temperature. The suggestion of Poulis and Massen (4) that the conductivity of sulfur might be measured by ac bridge techniques appears to be supported by the present data.

At lower temperatures (<170°C), electrical conduction in sulfur is thought to take place primarily by the electrophoresis of impurities, and the conductivity shows a direct correlation with the viscosity (1). With increasing temperature electronic conduction becomes predominant, and the conductivity shows a linear dependence of log κ upon $1/T$ (°K). This behavior is characteristic of, but not limited to, intrinsic semiconductors. The observed frequency dependence of the conductivity is in the opposite direction of that expected if electronic conduction occurs by a "hopping" mechanism (10). The activation energy does not show a dependence upon temperature which is characteristic of this type of mechanism, although such an effect may not be evident over the relatively short temperature range of the measurements.

Hyman (11, 12) has reported that the optical absorption edge for sulfur decreases from about 2.2 to 1.5 ev as the temperature increases from 210° to 430°C. Poulis and Massen have suggested that the poor correspondence between these values and the activation energy of 2.9 ev reported by Feher and Lutz is explained by assuming a more complicated model involving both conduction along polymeric molecules and that between separate molecules. The 1.62-ev activation energy from the present data is more in line with the results of the optical measurements, but it does not show the corresponding temperature dependence, possibly because of the relatively small temperature range of the data.

The conductivity data for P_4S_3, shown in Figure 2, show no obvious dependence upon frequency, and log κ varies linearly with $1/T$ (°K), corresponding to a rather low activation energy, 0.32 ev. At the higher temperatures, the conductivity data for P_4S_3 are in the same range as those for sulfur. It is clear from the data in Figure 2 that the conductivity of P_4S_3, like that of sulfur, is strongly enhanced by the presence of impurities, probably as a result of doping effects.

The dc current *vs.* time results showed a change of current amounting to only a few percent over time periods of thousands of seconds, resulting in dc conductivity values which were in agreement with the ac measurements, as discussed above. Closer examination of the small current change with time (passage through a minimum, with a gradual increase at long times for the higher applied voltages) leads to the following conclusions. The initial decrease of current is semilogarithmic with time, and corresponds to a capacitance of about $5\mu F/cm^2$, consistent with the charging of two electro-chemical double layers in series (one on each electrode). The subsequent increase of current at the higher applied voltages, coupled with the observation of a slight current instability, is consistent with the idea of a contribution of ionic impurities to the total current and a subsequent decomposition of the impurities (such as traces of phosphoric acid and/or sulfuric acid). These small contributions to the total current probably do not cause a significant error in the results.

In connection with the probable mechanism of electron transport in P_4S_3, x-ray diffraction studies (*13, 14*) have shown that solid P_4S_3 has a cage-like structure, and subsequent Raman work (*15*) indicates that this same structure exists in the liquid. Because of this structural characteristic of the P_4S_3 molecule, electronic conduction probably takes place between the individual molecules rather than along polymeric chains.

The data for mixtures of sulfur and P_4S_{10} shown in Figure 4 are subject to some uncertainty since they were corrected for the difference in conductivity between the lower purity sulfur used in these earlier experiments and that of the higher purity material used in the more recent work. Nevertheless, the results are believed to be sufficiently reliable to indicate that the addition of a few percent of phosphorus (as P_4S_{10}) to liquid sulfur causes a marked increase in its conductivity. The decrease in slope with increasing phosphorus content indicates that the effective energy gap for electronic conduction is narrowing, as might be expected. Addition of a few percent of phosphorus to sulfur results in a considerably higher conductivity than that of either sulfur or P_4S_3. It would be of interest to extend this work to include conductivity measurements of the other phosphorus–sulfur compounds (*e.g.*, P_4S_{10}, P_4S_7, P_4S_5) and intermediate compositions.

Acknowledgments

The authors are pleased to acknowledge the experimental assistance of L. W. Mishler, the encouragement and support of A. D. Tevebaugh, and the helpful discussions with V. A. Maroni and P. T. Cunningham. The purified P_4S_{10} used in this work was provided by V. A. Maroni.

Literature Cited

(1) Feher, F., Lutz, H. D., Z. Anorg. Allg. Chem. (1964) **333**, 216.
(2) Gordon, C. K., Phys. Rev. (1954) **95**, 306.
(3) Watanabe, O., Tamaki, S., Electrochim. Acta (1968) **13**, 11.
(4) Poulis, J. A., Massen, C. H., "Elemental Sulfur," p. 109, B. Meyer, Ed., Chap. 6, Interscience, New York, 1965.
(5) Curtis, H. J., Phys. Rev. (1932) **41**, 386.
(6) Kraus, C. A., Johnson, E. W., J. Phys. Chem. (1928) **32**, 1287.
(7) Forthmann, R., Schneider, A., Naturwissenschaften (1965) **52**, 390.
(8) Jones, G. P., Bradshaw, B. C., J. Amer. Chem. Soc. (1933) **55**, 1780.
(9) Johnson, C. E., Foster, M. S., Kyle, M. L., Nucl. Appl. (1967) **3**, 563.
(10) Mott, N., Advan. Phys. (1967) **16**, 49.
(11) Hyman, R. A., Nuovo Cimento (1955), Suppl. 2, Series X, 754.
(12) Hyman, R. A., Proc. Phys. Soc. (1956) **B69**, 1085.
(13) Houten, S., Vox, A., Wregers, G. A., Rec. Trav. Chim. (1955) **74**, 1167.
(14) Leung, Y. C., Waser, J., Roberts, L. R., Chem. Ind. (London) (1955) 948.
(15) Maroni, V. A., Hathaway, E. J., Cairns, E. J., "Chemical Engineering Division Annual Report—1969," p. 138, Argonne National Laboratory, ANL-7675 (April 1970).

RECEIVED March 5, 1971. Work performed under the auspices of the U. S. Atomic Energy Commission.

15

Chemical–Mechanical Alteration of Elemental Sulfur

JOHN M. DALE

Southwest Research Institute, San Antonio, Tex. 78213

Elemental sulfur is altered by chemical, allotropic, or me-chanical means. Of the chemical modifiers the Thiokol family of additives are some of the most effective modifiers available. By control of the time–temperature history of sulfur in the liquid and solid phases, it is possible to control the allotropic modification of sulfur and its mechan-ical properties. Mechanical modification of sulfur with non-chemically reactive filler materials such as aggregates and fibers yields materials with properties different from those of the parent materials.

If elemental sulfur is used as a mechanical or structural material, three characteristics of the material must be considered. These character-istics are the low tensile strength of sulfur, the strain rate sensitivity of sulfur, and the fact that, as the name brimstone implies, sulfur burns. The order of importance of these characteristics is of no significance other than it relates to a specific application. A coating or structural element in a building that burns is a serious matter. A road subsurface membrane that burns is of little significance.

Chemical Modification

When drawn into thin filaments, sulfur exhibits like many other materials in this form outstanding strength properties (*1, 2, 3*). If quench cooled from above the transition temperature, it exhibits its plastic–elastic behavior. Although great effort has been expended searching for additives and techniques to stabilize plastic–elastic sulfur, little success has been achieved. There are only a few additives that stabilize these characteristics for even short periods of time. They include the elements:

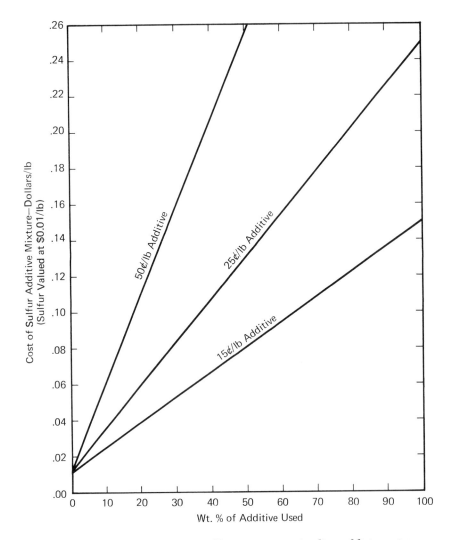

Figure 1. Influence of cost of additive on cost of sulfur additive mixture

arsenic (4), phosphorus (5, 6, 7, 8), selenium (5), tellurium (5), and thallium (4). There are a variety of compounds that are more effective than the above elements, but they are expensive generally, require high levels of addition, and do not impart permanent changes.

The chemical routes to altering the properties of sulfur have been presented by Barnes (9). These techniques often result in lowering the compressive and flexural strength of the sulfur, imparting color and odor to the sulfur from by-products generated, and raising the cost of the resultant materials by the cost of the additives and the labor required to introduce them. When the available modifiers are used in the amounts needed to achieve the desired physical property changes, the cost of the product often overcomes the low cost advantage of sulfur which justified starting with sulfur in the first place. It is not unusual to exceed quickly the value of the sulfur with the value of the additive. Figure 1 shows the influence of several different cost additives at different levels of addition.

The specific gravity of sulfur is almost 2, whereas many of the materials that sulfur seeks to compete with such as plastics have specific gravities in the range of 0.9 to 1.4. Usually the sulfur material must equate with these other materials on a volume basis rather than on a weight basis for structural reasons. Figure 2 shows how the costs are altered upward when this factor is considered.

If, for example, one had a $0.50 per pound chemical modifier for sulfur that would improve the tensile strength of sulfur at a 30% by weight addition from 150 psi up to 2,000 psi, then this composition of $0.16 per pound might be looked upon at first as a possible substitute for $0.22 per pound polyethylene. If the material were considered for use in the injection molding of some item then structural factors would be involved, and it is likely that equal volumes of the materials would be required. The net result is that the sulfur mixture would cost about $0.24 which is more than the polyethylene, and there would be no incentive for its use.

In screening chemical modifiers for sulfur one looks for more than transitory improvements in the physical properties (particularly tensile and impact strength), the burning characteristics, the color, and the odor. Changes in viscosity, surface tension, and supercooling tendencies are considered secondary areas.

Of the various chemical modifiers for sulfur which we have examined over the years we must single out the Thiokol family of additives as some of the most effective sulfur modifiers available (*10*). Unfortunately they are expensive, but their effects are long term. One of the most deceptive of the sulfur additives is styrene monomer (*11*) which is attractive because of its low cost. One can obtain an unusually wide variation of properties in sulfur–styrene mixtures by controlling the degree of reaction. Unfortunately these property modifications are transient, so much so that we use it infrequently and for only some special purpose. The transient characteristics of styrene–sulfur mixtures are attributed to the attack of the polystyryl radicals on the sulfur–sulfur bonds and the rapid opening of the sulfur rings.

This effect of styrene on sulfur may have an important role in one of the special uses found for styrene. Styrene is used advantageously with sulfur in rendering sulfur nonburning in accordance with ASTM Test Methods D635 (*12*). It was found that a number of materials such

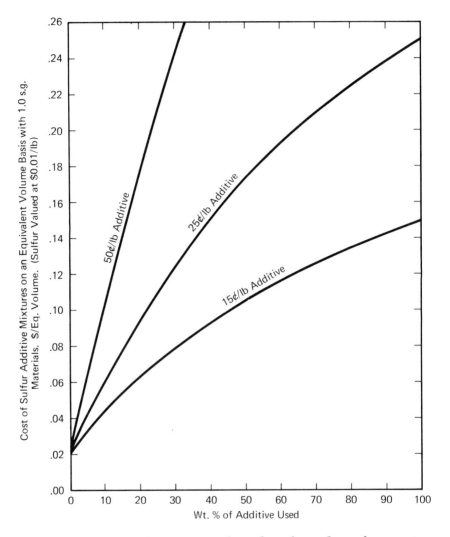

*Figure 2.   Cost of sulfur mixtures when adjusted to reflect volume equiva-
lence with conventional plastics*

as maleic acid, tricresyl phosphate, and Chlorowax 70S act as fire re-
tardants for sulfur when in the presence of styrene. The "nonburning"
definition is qualified as being by the ASTM Test Method D635; in applications the sulfur formulations are generally in contact with other materials which often negate many of the hard won improvements. Admix-
tures with glass fibers are probably the most difficult to fire retard because

the fiber bundles act as small wicks and greatly stimulate the combustion process. One promising approach that we have examined is to incorporate an inert filler or some material that crosslinks with sulfur and chars at high temperature. When the material catches fire, only the surface burns, forming a char or intumescent coating which prevents oxygen from getting to the underlying material, and the composition is therefore self extinguishing. While progress has been made in fire retarding sulfur, a great deal remains to be done, and advances in this area will contribute to the use of modifications of elemental sulfur in various applications. Plastics, many of whose components have heats of combustion of from 16,000 to 20,000 btu's per pound, have been fire retarded. Sulfur, with its low 4,000 btu's per pound heat of combustion, possibly can be made completely fire resistant with further development work.

Allotropic Modification

The mechanical properties of sulfur can be altered also by control of the allotropic crystalline modifications of sulfur through time–temperature control of the liquid and the solid. Many values for the tensile strength of sulfur appear in the literature; they range from 160 psi (*1, 13*) for bulk material to 14,000 psi for sulfur threads. The test values vary greatly with the method of test—i.e., the size of the specimen, the load rate, and the polymeric sulfur content. The strength of sulfur can be correlated with the polymeric sulfur content. In some of our recent unpublished work it appears that the strength of sulfur is not a direct line function of the polymeric sulfur content; the strength goes up, peaks, and then declines as the polymeric sulfur content of the sulfur increases. This suggests that the polymeric sulfur reinforces the orthorhombic sulfur. It is often possible to control allotropic modifications, and therefore strength, on a commercial basis to effect such things as a reduction in fines and improvement in uniformity for a material to be transported, or the reduction of grinding costs for a material requiring grinding.

Mechanical Modification

Some of the more significant modifications of sulfur have been achieved by mechanical modification with nonchemical reactive filler materials such as aggregate and fibers. Sulfur-fine aggregate and sulfur-fine and coarse aggregate mixtures have been developed with compressive strengths of up to 10,700 psi (*14, 15*). In most instances mixtures of sulfur and aggregate materials increase in strength with an increase in the sulfur content up to a point beyond which the strength declines. For most sulfur-aggregate mixtures the strength peaks at a sulfur content of between 16 and 25% by weight. It does not follow necessarily that high strength aggregates mixed with sulfur will give greater strengths than low strength aggregates mixed with sulfur; wetting characteristics between the sulfur and the aggregate and the cleavage form in the aggregate materials often are overriding. Sulfur aggregate mixtures that have no chemical modifying additives are much less resistant to freeze–thaw cycling than mixtures containing modifiers.

Characteristically, two-phase materials are made up of two different substances with contrasting properties of strength and elasticity. Under an applied force a low modulus substance stretches and deforms and distributes the stress to the high strength component. Sulfur has a modulus in the vicinity of 1×10^6 psi so glass and asbestos (with moduli of 10×10^6 and 24×10^6, respectively) should theoretically and in practice do combine well with sulfur (16). Glass fibers can be used to reinforce sulfur so that flexural strengths as great as 7,000 psi can be obtained (17). However, glass fibers obtained commercially generally have sizing and binder materials on them that are made to be compatible with other resins and not sulfur. Therefore, it is necessary to remove these treatments or apply other treatments to the glass if intimate wetting of the glass by the sulfur is to be achieved and maximum strengths obtained. Wherever glass fibers are used with sulfur formulations, the glass should be imbedded well in the sulfur and coated over by the sulfur so that exposed glass fibers can not focus sunlight and initiate combustion.

Another basically mechanical modification of sulfur requiring a great deal of chemical assistance is the production of sulfur foam. Elemental sulfur on melting has a viscosity and surface tension much like that of distilled water which is certainly not conducive to froth formation. By the addition of various additives it is possible to change the viscosity, surface tension, and film forming characteristics of sulfur so that it can be altered mechanically by the inclusion of a gas to form a molten froth which cools to a rigid foam (18). An expansion ratio of 10 to 1 is achieved easily, and the properties of the resultant foams are greatly different from those of the parent material. They can be produced with a crushing stress that remains constant up through as much as 80% of their deformation, a desirable characteristic of shock isolation materials. Sulfur foams can be produced with an essentially closed cell characteristic which resists the passage of liquids and vapors. These same foam materials have low thermal conductivities and are quite attractive as structural thermal insulation material.

In the laboratory we have combined many and varied types of materials with sulfur. Although we are just learning about the chemical–mechanical modifications of sulfur, we are beginning to be able to define the limits of present technology in this area.

Literature Cited

(1) Sakurada, K., Erbing, H., *Kolloid-Z.* (1935) **72**, 129.
(2) Trillat, J. J., Forestier, H., *Compt. Rend.* (1931) **192**, 559.
(3) Trillat, J. J., Forestier, H., *Bull. Soc. Chim.* (1932) **51**, 248.
(4) Hamor, W. A., Duecker, W. W., U.S. Patent **1,981,232** (Nov. 20, 1934).
(5) Ellis, C., "The Chemistry of Synthetic Resins," p. 45, Vol. 2, Reinhold, New York, 1935.
(6) Hamor, W. A., Duecker, W. W., U.S. Patent **1,959,026** (May, 1934).
(7) Specker, H., Z. *Angew. Chem.* (1949) **61**, 439.
(8) Specker, H., Z. *Anorg. Chem.* (1950) **261**, 116.
(9) Meyer, Beat, "Elemental Sulfur," pp. 161–178, Interscience, New York, 1965.
(10) Placek, C., "Polysulfide Manufacture," pp. 2–38, Chemical Process Review No. 5, Noyes Data Corporation, 1970.
(11) Pryor, W. A., "Mechanisms of Sulfur Reactions," p. 10, McGraw-Hill, New York, 1962.
(12) Dale, J. M., Ludwig, A. C., "Fire Retarding Elemental Sulphur," *J. of Mater.* (March 1967) **2**, 1, 131–145.
(13) Dale, J. M., Ludwig, A. C., "Mechanical Properties of Sulphur Allotropes," *Mater. Res. Stand.* (August, 1965) **5**, 8, 411–417.
(14) Dale, J. M., Ludwig, A. C., "Sulphur-Aggregate Concrete," *Civil Eng.* (December 1967) **37**, 12.
(15) Crow, Lester J., Bates, Robert C., "Strength of Sulphur-Basalt Concretes," *U.S. Bur. Mines Rep. Invest.* (March, 1970) **7349**, 1–21.
(16) Dale, J. M., "Sulphur-Fibre Coatings," *Sulphur Inst. J.* (September, 1965) 11–13.
(17) Dale, J. M., Ludwig, A. C., "Reinforcement of Elemental Sulphur," *Sulphur Inst. J.* (Summer, 1969) **5**, 2, 2–4.
(18) Dale, J. M., Ludwig, A. C., "Rigid Sulphur Foam," *Sulphur Inst. J.* (Autumn, 1966) **2**, 3, 6–8.

RECEIVED March 5, 1971.

16

Some Potential Applications of Sulfur

HAROLD L. FIKE

The Sulphur Institute, Washington, D. C. 20006

Sulfur is a low cost, high purity, readily available material. It exists in many allotropic modifications with interesting physical and chemical properties. In 1970 world consumption of sulfur in all forms totaled 38 million tons. However, only a small percentage of the sulfur consumed depended upon or benefited from its high purity and unique chemical and physical properties. Sulfurization of asphalt, sulfur impregnation of ceramics and paper, and the use of sulfur in construction are a few potential applications which could exploit sulfur's unique properties. The present status and additional research needed to commercialize these potential applications are discussed.

Throughout history continued references are made concerning the benefits of sulfur. The development of efficient processes for manufacturing sulfuric acid made sulfur the workhorse of the early chemical industry. It has held this position for several centuries, and in 1970 world consumption of sulfur in all forms totaled 38 million metric tons. Principal end uses include the manufacture of fertilizer, fabrics, paper, steel, petroleum, rubber, and thousands of economically important and technically necessary compounds.

Because of its long history and economic importance, it is assumed that little remains to be learned concerning the chemistry of sulfur. However, this statement is not true, and even its melting point is still subject to debate (1). Therefore, the development and commercialization of promising new uses for sulfur are delayed because of the lack of information on this element's properties and behavior.

Additional research on sulfur and its compounds is needed in several fields. Our knowledge of biologically important sulfur compounds such as the essential amino acids and sulfur-containing pharmaceuticals and insecticides is not sufficient.

Few argue that we have much to learn about the Thiokols and other polysulfide elastomers, the polysulfones, which have such interesting properties as the engineering plastics, and the episulfides. The chemistry of pyrites, carbon disulfide, sulfur dioxide, dimethyl sulfoxide, and many other compounds merit separate reviews.

However, only the chemical and physical properties of elemental sulfur which can contribute to the development of new uses for this material will be discussed. Even in this narrow subject area the reader must consult the references for more than a superficial discussion of many applications.

The need for additional research is related to sulfur's complexity. It exists in several allotropic forms which differ in their physical and chemical properties. The forms have different compressive and flexural strengths and differ in solubility in various solvents (2, 3). Some allotropes are brittle while others exhibit viscoelastic behavior (4). Unfortunately, the only allotrope stable under standard conditions of temperature and pressure appears to have less useful properties than many of the metastable forms. This situation makes identifying the metastable allotropes and devising means of preparing them desirable. Schmidt (5) and others are making notable advances in this area. It is important to find materials which encourage the formation and stabilization of the more useful forms. Barnes (6) points out that selenium, diolefins, and other difunctional materials lead to the formation of linear sulfur polymers. Other additives such as arsenic, phosphorus, and the tri- and higher functional organic compounds result in crosslinked polymers. However, impurities, the amount of additive (7), and the time–temperature history of the sulfur are also important variables.

Sulfurized Asphalts

Asphaltic pavements, roofing materials, and coatings deteriorate from the effects of light, water, heat, and air (8, 9). In addition to photochemical and chemical reactions, bacterial action (10) also contributes to the eventual failure of these materials. Asphalt materials become brittle at low temperature, and severe winters cause extensive damage to pavements. Applications of the materials at low ambient temperatures are also a problem and probably reduce durability.

Too much has been written on sulfurized asphalts to permit complete coverage in this paper. However, some of the more pertinent developments on the subject are summarized below, and more comprehensive discussions are available (8, 11, 12, 13, 14).

The need to improve asphaltic materials was recognized early, and the process for treating asphalts with sulfur was introduced commercially over 100 years ago (8). A. G. Day and, subsequently, J. A. Dubbs showed that Pennsylvania and Ohio residuums heated with 20–25% sulfur were more resistant to weathering and more durable than normal asphalt, especially at low temperatures. Sulfur addition appeared to flatten the viscosity–temperature curve. Increasing the viscosity at high temperatures and lowering it at lower temperatures is desirable in most paving and coating uses. These sulfurized asphalts were produced under the trade name Pittsburgh Flux. During the latter part of the 19th century, air-blown petroleum asphalts were developed. The air-blown asphalts had some properties of the sulfurized material, could be produced at lower cost, and eventually replaced Pittsburgh Flux.

The variable nature of the raw materials used to manufacture blown asphalts and the lack of quality control in the manufacturing method limited the early growth in demand for these products. Kronstein (15) and others (16) describe techniques for improving these materials.

During the past half century many papers have been published on the merits of sulfurized asphalts and techniques for improving the products. Early investigators were intrigued by the similarity between sulfurization of asphalts and the vulcanization of rubber. Unfortunately rubber chemistry is complex, and recognition of this similarity did not immediately advance the state of the art. However, the similarities are striking. The constituents of natural rubber vary slightly from tree to tree and considerably among trees grown in different geographical areas. These variations led to many failures in the early days of rubber vulcanization. However, in rubber vulcanization the larger and more obvious economic incentive and the need for durable elastomers was so great that technical problems were overcome (17).

Rubber compounders also learned to live with a certain number of failures. Accelerators and other additives which reduce curing time and prevent sulfur blooming were developed. The possible effects of the various allotropic forms of sulfur in rubber have been reported (18, 19, 20). Although too costly to have gained wide commercial acceptance, organic sulfur compounds make excellent vulcanizing agents for rubber (21). Related compounds cure bituminous mixtures (22).

Natural and synthetic rubbers have been studied as additives to alter the viscosity, ductility, and flow properties of road building asphalts. Welborn and Babashak (23) added natural rubber and sulfur to asphalt and reported improved blending conditions, improved stability, increased toughness, and low temperature ductility.

Bacon and Bencowitz (24) concluded that burning the surface of sulfurized asphalts induced an additional chemical reaction which resulted in improved durability.

Sommer (25) claimed that the adhesivity of asphalt to aggregate is improved by adding elemental sulfur and amines having 12–14 carbon atoms. Hughes and Stine (26) indicated that the adhesivity of asphalt to aggregate is improved by adding sulfur and phosphorus sulfides.

Numerous papers contradict and appear to refute the above claims. However, an examination of these papers indicates that they are less contradictory than they first appear. Many investigators considered coal tars and a variety of different petroleum fractions produced by different processes as equivalents. Others compared the natural sulfur content of petroleum with materials to which elemental sulfur had been added. However, not all of the contradictory evidence was explained, and in some cases investigators obtained different results even though they were attempting to duplicate each other's efforts. Cases where the same investigator was unable to reproduce data are known also.

Factors which probably have contributed to the disagreements among research workers and the disparity in the results obtained include the point made at the beginning of this paper—sulfur is a more complex element than most of the early investigators realized. Few of the published papers recognize that sulfur exists in many different allotropic forms and that the different allotropes have different solubilities and different reaction rates. It is assumed that the different allotropes yield different asphalt materials in the same way that they produce different elastomers (18).

The allotropic composition of sulfur is altered by temperature and by the presence of hydrogen sulfide and other impurities (27) which are present in petroleum asphalt. Quarles Van Ufford (28) in his text presents a comprehensive and precise reporting of his work. However, even here questions are raised concerning the specific time–temperature conditions which might have affected the allotropic composition of the sulfur used in his tests. Few other investigators recognized allotropic composition of the sulfur as a variable, and this failure contributed to some of the early confusion and disagreements.

Another gap in our knowledge is the difficulty of relating the properties of asphalts and specific test results to performance in each application (29). Penetration, viscosity, ductility, and elasticity properties are related to performance. However, the problem of correlating results from many of these empirical tests with the development of additives to increase asphalt's durability has created difficulties in interpreting the results (30, 31, 32, 33, 34, 35).

The lack of appreciation for the different properties of bitumens being tested created part of the problem. The complexity of bituminous materials makes it difficult to identify reaction products. The failure in the literature to specify the nature of the bituminous material used reduces the value of much published data.

**Table I. Addition of Sulfur (7.5%) and Carbon (7.5%)
to Paving Asphalt** [a]

	Control	*Treated*
Surface texture	average	much above average
Crack resistance	average	much above average
Rut resistance	average	above average
Abrasive wear	average	above average

[a] Service ratings after 2.5 million wheel passages.

Despite the conflicting and sometimes unreliable data, one conclusion appears inescapable—on many occasions the addition of sulfur to asphalt has produced materials with improved strengths and durability as proved in usage. Various organizations have prepared floors and road sections which demonstrate forcefully that beneficial results have been obtained even if they were not always reproducible. Unfortunately, little of this work is published. However, several papers (36, 37) do describe the benefits obtained under separate sets of conditions. Tables I–III present data which quantify some of these benefits.

Improved knowledge about sulfur chemistry, the strides being made in developing better testing procedures, and the increased understanding of the nature and properties of petroleum asphalts encourage the belief that future studies on this subject will be rewarding (38, 39). Extension of early work which theorized on the mechanism of sulfurization would yield further insight. The present economic situation creates a favorable incentive to reopen this investigation. Asphalt is in short supply, and the prices are rising. Sulfur, which has been in short supply during much of the past decade, is now readily available. Thus, sulfur added to asphalts is justified as a low cost diluent rather than as an expensive additive.

Sulfur Impregnation of Ceramic and Cement Materials

Impregnation of bonded abrasive grinding wheels with sulfur improves their strength and acts as a lubricant and coolant during grinding operations. Sulfur impregnated wheels are well suited for grinding tough materials such as stainless steel cutlery and materials made of brass,

**Table II. Addition of Sulfur (7.5%) and Carbon (7.5%)
to Paving Asphalt**

	Control	*Treated*
Compressive strength, psi	116	650
Tensile strength, psi	2	31

bronze, and nickel. The impregnated wheels cut faster and prevent the welding of metal chips. In more difficult jobs such as gear grinding and surface grinding, the sulfur impregnated wheels have four to eight times the life of high quality, non-impregnated wheels.

The usual impregnating procedure is to immerse the abrasive in liquid sulfur until the pores are filled. The use of vacuum systems to remove air from the body prior to immersion in the liquid sulfur speeds greatly the rate of impregnation. Inherent in both methods are such problems as: sulfur runs out of the body after it is removed from the liquid or the sulfur solidifies to form a skin on outer portions of the body. Both can lead to uneven draining which produces imbalanced wheels. Slow withdrawal from the bath together with controlled heating are suggested remedies (40). Impregnation through capillary action is not reported in the literature, but it appears to merit consideration.

Most commercial forms of sulfur have been used for impregnation. Data relating strength and durability to the allotropic form, bath temperature, and cooling rate after removal from the impregnating tank have not been published although some of this work has been done. The allotropic composition is an important variable in the performance of these abrasives. Manufacturers of the impregnated abrasives report that increasing the rate of production and minimizing the odor of the sulfur or sulfur compounds reduces the cost, improves the aesthetics, and encourages use of the method.

Sulfur has been used also to impregnate ceramic tile. Sulfur impregnation reduces water absorption and makes the tile frost resistant when used on exterior surfaces such as floors, roofs, or entrances to buildings. It also makes tile more resistant to staining by grouts and mortar and improves the bond strength when using many conventional adhesives. Impregnation improves also impact and compressive strength which is important in floor tile used in heavy traffic areas.

The Tile Council of America developed a simple impregnation method (41). Ceramic tiles are preheated to 150°C under a vacuum of about 30 inches of mercury. The tiles are immersed in liquid sulfur, and impregnation is completed in several minutes. Kakos and Fitzgerald (41) report that orthorhombic sulfur allotropes are preferred. Amorphous sulfur is claimed to produce tile with inferior strength and causes diffi-

Table III.   Addition of Sulfur to Paving Asphalt

Sulfur, %	Asphalt, %	Aggregate, %	Stability, lbs	Bearing Capacity, psi
0	7.25	92.75	995	109
5.5	5.5	89.0	2905	551
10.0	4.5	85.5	4575	640

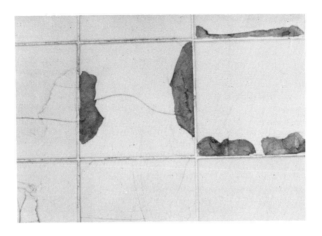

Figure 1. Wall tiles in an exterior location. After five years, the standard tiles (top) have deteriorated badly whereas the impregnated tiles (bottom) show no damage.

culty in removing excess sulfur and cleaning the impregnated tile. Allotropic composition is controlled by holding the sulfur bath below 159°C.

Figure 1 depicts two arrays of glazed wall tile which were installed on an exterior test wall. After five years the standard tiles shown in the top photograph have deteriorated badly. The impregnation prevents the tiles from absorbing water, and, as can be seen in the bottom photograph, the sulfurized tiles essentially are undamaged. Table IV compares the quality of standard and impregnated tiles and shows the improvements resulting from sulfur treatments.

Floor panel tests provide a practical demonstration of the improved strength resulting from sulfur impregnation. In one test rubber and steel wheels were run over a panel consisting of a matte glazed wall tile and a crystalline glazed wall tile which is used also on light duty floors. Half of each type of tile had been sulfur treated. After five hours of testing the sulfur treated, matte glazed tile showed less damage than either of the non-sulfurized tiles. The sulfur impregnated crystalline glazed tile was undamaged.

The impregnation process takes place on finished tile which permits the technique to be incorporated into a manufacturing plant without disrupting the normal operating procedures. Eliminating the need for a vacuum system increases the impregnation time but probably improves the manufacturing economics. Research on the time and temperature conditions and on additives which permit control of the allotropic composition could maximize strength or other desirable qualities (2).

Sulfur Impregnation of Paper

Investigators at the Institute of Paper Chemistry developed inexpensive sulfur impregnated paperboard houses (42). The houses were designed to perform as mass produced, portable shelters capable of withstanding all weather hazards for periods exceeding one year.

By designing the houses as frameless structures, the investigators minimized the need for wood and metal. However, in frameless houses the wall panels must serve as the main load-bearing structural support, and standard paperboard materials are too weak to be considered. Paperboard impregnated with asphalts or other organic materials creeps when subjected to sustained loads. Creep was too severe with asphalt to justify more than preliminary testing. Of the other materials which might impart sufficient strength, all except sulfur were too costly to merit serious consideration.

Since sulfur, without additives, has limited strength and becomes brittle as it reverts to the orthorhombic crystalline form, the investigators expected major problems with sulfur impregnation. For example, sawing and nailing operations might lead to fissures and cracks which would be

Table IV.   Sulfur Impregnation of Ceramic Tile

	Untreated	*S-Treated*	*Change*
Water absorption	21.3%	1.3%	−94%
Tensile strength	167 lbs	331 lbs	98%
Compressive strength	7800 psi	25,250 psi	223%

unacceptable. However, few of these problems arose during the investigation, and the qualities of the sulfurized paperboard were good. The investigators believe that the paper fibers may reinforce the sulfur, impart impact resistance, and improve compressive strength. Subsequent studies demonstrate the benefits of fiber reinforcement (7, 43). Other work (44) has shown that rosin and some terpene compounds stabilize sulfur in various metastable forms. Thus chemicals derived from the paperboard might enhance the properties of sulfur. Nails are driven through the material without cracking it, and sawing operations are accomplished with ease.

The investigators expected that retaining adequate strength under hot, humid conditions might be a major problem which could prevent the use of sulfurized paperboard as a housing material. Although the sulfurized paperboard does weaken under extreme conditions, the loss plateaus at a satisfactory value, and the board meets the necessary specifications.

Reduced strength after immersion in water and the fire hazard also were considered serious obstacles. However, surface treatment with a variety of available paints waterproofed the paperboard. Many of these paints also imparted fire resistance and made the material self-extinguishing in laboratory tests.

In addition to the laboratory testing, houses constructed of one-inch thick panels performed well in outdoor tests. As expected, the sulfurized paperboard was a good thermal insulating material. After two years' service the houses were sound and strong. During this period the houses withstood intense winds and were strong enough to withstand large rocks being thrown at them.

The impregnation method is simple (42). Dry paperboard is immersed in sulfur held at 135°–140°C. Impregnation of 4 × 8 foot panels requires only 15 minutes. Removal of the panel from the sulfur bath requires that the panels be supported until the sulfur solidifies. The panel is held at a sharp angle to permit quick draining of the excess sulfur. Impregnation temperatures above 140°C weaken the fibers. Impregnation time is increased significantly above the transition temperature (159°C) because of increasing viscosity.

A different technique (45) was developed for manufacturing sulfurized corrugated container board suitable for self-sustaining shipping containers. Impregnation is performed by applying the sulfur to the corrugating medium and ironing or hot rolling it into the board. The ironing process conditions the fibers and is carried out on one or both sides of the medium. The container board is prepared by applying adhesive to the crests of the corrugated material and pressing the liners into place in the normal manner. The sulfur content of the corrugating material exceeds

50% but accounts for only about 15% of the weight of the finished container board.

Sulfurized container board is shock resistant and maintains its strength under hot, humid conditions which cause normal box-board to lose over half its strength.

Sulfur Coatings

Sulfur's thermoplastic property and its resistance to chemical attack led to its use as a decorative and functional coating (*46, 47*). Unless additives are used to prevent reversion to a crystalline allotrope, the coatings develop pinpoint holes and lose strength when used as a jointing material. When pure sulfur was used to protect concrete for storing hot saline solutions, it failed in less than one year (*48*).

The early sulfur cements were used to join clay sewer pipe, clay tile, cast iron pipe, paving brick, acid reaction vessels, and for similar applications. The composition of the materials varied but were mostly sulfur containing 1–10% of an organic polysulfide. Silica, carbon black, and wood rosin also were used frequently. However, the compositions varied considerably, and one promising material contained 60% air blown asphalt and 40% sulfur (*49*).

Sulfur cements are resistant to attack from water, hydrochloric, nitric, hydrofluoric, sulfuric, phosphoric, and most organic acids. When tested as linings for storing acids at concentrations of 5–90% and at temperatures up to 90°C, the sulfur coatings were unaffected generally after several years of service.

In protecting against alkaline solutions, the sulfur coatings are less effective. At room temperatures the cements give adequate protection against saturated lime solutions and intermittent use of dilute caustic. However, strongly alkaline materials and even dilute alkalies at 80°C will attack sulfur cements. Dichromates and other strong oxidizing agents also cause sulfur coatings to disintegate; hence, sulfur coatings are not recommended for these materials.

The common inorganic salts or organic materials have little effect on sulfur coatings unless they are solvents for sulfur. Thus, carbon disulfide, the bisulfites, and aromatic compounds such as xylenes cannot be stored in sulfur lined vessels.

Studies on the use of sulfur as a highway marking material were initiated in the 1950's (*44*), and during the past decade this idea was tested throughout the world. The thermoplastic nature of sulfur makes it an ideal marking material. It solidifies usually in less than one minute and returns the highway to service with a minimum of delay. Most marking paints used today are slow drying and require protective devices for several hours after application. The quick setting sulfur paints might reduce the number of accidents which occur during striping operations.

The composition of sulfur paints varies, but a typical formulation is shown below:

Constituents	Percent by Weight
Sulfur	70–95
Plasticizer	5–8
Pigments or whitening	0–25
Masking agents, chain terminating compounds, etc.	0–5

Plasticizers which have been used successfully include polyethylene tetrasulfide, polyethylene diacetate tetrasulfide, dimercaptobutane diallyl tetrasulfide, and the (mercaptoethyl) cyclohexanethiols (6, 50, 51, 52).

The development of yellow and other colors for highway marking paints was accomplished easily. However, a white paint which meets color and reflectance specifications is difficult to prepare. The absorption characteristics and high refractive index of elemental sulfur present formidable problems. The simple addition of titanium dioxide and other white pigments does not produce a white paint, and masking agents or other additives usually must be incorporated.

Holding sulfur which contains large amounts of pigments at elevated temperatures introduces another problem. The viscosity increases, and the paint becomes unsprayable. Kane (53) recommends the use of chain terminating compounds such as the mercaptoethanols to maintain the viscosity at low levels. Some of these compounds are effective even when used at concentrations as low as 0.1%.

The sulfur based marking paints have shown outstanding wear characteristics on asphalt and concrete pavements (54). They performed better than standard paints in all climates and weather when tested in residential communities, rural areas, and in parking lots, as well as on busy highways and in major metropolitan areas. Some of the advantages of sulfur paints are summarized in Table V. Relatively sophisticated application equipment which automatically dispenses reflecting beads was developed and tested in North America and Europe.

Table V.   Advantages of Sulfur Marking Paints
over Conventional Paints

Greater durability
Very short set time
Excellent reflecting bead retention
Good weathering properties
Easily stored—no drums to turn

Sulfur Surface Bond Construction

In the early 1960's investigators developed sulfur coatings which could replace mortar in masonry construction (55). Bricks, concrete, cinder, or clay blocks are stacked one upon the other until the desired wall configuration is achieved. The sulfur coating is brushed or sprayed on both exterior surfaces. The sulfur does not need to penetrate between the blocks, and the need for mortar is eliminated. The coating dries almost immediately and forms a strong, hard, and insulative surface.

The need for low cost building techniques is large and increasing rapidly in most areas of the world (56). For reasons of climate and insect attack, wood buildings are impractical in many areas. In addition, the cost of wood and the high cost of the skilled labor needed to construct wooden shelters precludes its use under certain conditions. This situation has led many governments (57, 58) and private companies to investigate new building techniques and materials. The sulfur surface bond construction method has many of the qualities sought by these groups. For example, when compared with conventional mortar masonry construction:

(1) The sulfur method has much greater flexural strength, racking strength, and tensile or bond strength. Impact and durability tests also indicate superior performance.

(2) Relatively unskilled labor can be used in sulfur construction. Until the coating is applied, the blocks can be adjusted and plumbed as often as necessary to insure that the wall is straight.

(3) The bond sets within a few minutes and does not require curing to yield good bonds or to prevent cracking. Construction of roofs and rafters can be started without the 12–24 hour curing period.

(4) The coating can serve as a waterproofing and decorative paint.

As originally developed, the sulfur coating contained about 5% glass fiber to improve its strength and 5% of an organic polysulfide material to control viscosity and prevent crystallization. A small warehouse and a number of wall sections were built using the technique. The following disadvantages were observed:

(1) Brushing the hot sulfur with conventional paint brushes is time consuming.

(2) Sulfur is flammable, and the use of fibers increases this hazard.

(3) Although used in small amounts, the cost of the additives is 3–4 times the cost of the sulfur. This factor adversely affects the cost advantage.

(4) Building codes and trade union regulations present formidable barriers to the use of the sulfur construction method as well as other new building techniques.

The United Nations (U.N.) has sponsored projects at the Civil Engineering Department of Columbia University (59) and at Southwest Research Institute (58). U.N. personnel worked with these groups and others in improving and demonstrating the feasibility of sulfur surface bond construction. The improvements overcame many of the disadvantages listed above.

It appears that satisfactory strength can be achieved even though fibers are eliminated from the formula (59). Although the non-fiber containing formulations have lower strength than those containing fibers, data suggest that the former are still stronger than conventional mortar. Another important finding was that the thinner (0.04 inch) coatings impart greater strength than the heavier (0.10 inch) coatings. If these data can be substantiated, it will be of particular significance in that:

(1) The cost of the coating will be reduced even further.

(2) The development of spray equipment to replace brushing will be simplified.

(3) The fire hazard is reduced. Additives which make the non-fiber coatings self-extinguishing have been developed (60, 61).

To overcome the legal and social barriers, investigators will probably emphasize:

(1) Construction of storage bins and buildings for use by farm animals

(2) Construction of low-cost, single-story housing in developing nations

(3) That sulfur must be made non-burning rather than self-extinguishing.

Other Applications

With small amounts of additives, sulfur is turned into a rigid foam with many interesting properties (60). The foam is an excellent thermal insulator, weighs as little as 10 lbs/cu ft, and has compressive strengths of 50–500 psi. The compressive strength remains relatively constant for strains up to 50%. Low water absorption, good bonding to slabs poured earlier, and low shrinkage are also attractive characteristics of the sulfur foams.

Although surface active agents, foam stabilizers, blowing agents, and viscosity control compounds are used as additives, the foams are primarily (90%) sulfur.

Potential applications include use as an insulative base for foundations and highways, in the manufacture of foam core panels, and as a general filler or insulation material. Sulfur foams can be foamed in place, and this property enhances their value in many applications.

Sulfur is an excellent bonding agent for limestone, sand, gravel, and other common aggregates (62, 63). Sulfur aggregate concretes with strengths greater than 10,000 psi have been developed (64). Its high early strength and the fact that it does not require placement within a specified length of time are important advantages when comparing it with conventional concretes. This material appears to be particularly useful in bonding anchor bolts and patching asphalt or concrete surfaces (65, 66). Recently efforts have been directed toward using sulfur aggregates as structural materials.

The addition of sulfur to polyolefins and other polymers has been suggested for making agricultural mulches biodegradable. Incorporation of sulfur into plastic bottles and containers might make these materials self-destructing and minimize solids disposal problems (6). The use of sulfur as a biodegradable coating to control the release of nutrients contained in fertilizers is undergoing study (6).

Most of these uses, as well as the applications mentioned earlier, could be developed faster and at lower cost if more were known about elemental sulfur. Although progress is being made and reported, Meyer's observation of six years ago is still valid: "Many observations on elemental sulfur are unexplained, or poorly documented, and many conclusions have been contradicted. We are still far from the day when a truly thorough description of the element can be given" (60).

Research Needs

Sulfur research is increasing and will benefit the development of new uses for sulfur. Evidence which increases the accuracy of data or explains some of the contradictions reported in the literature is considered progress.

The need for additional research is obvious since we are attempting still to determine the melting point of sulfur. Thackray (1) reported recently that the melting points of four allotropes of cyclooctasulfur are 120.14°, 115.11°, 108.60°, and 106.0°C which are different from accepted values. How better to illustrate the need for research than to find we still disagree on its melting point?

On the other hand, Vezzoli *et al.* (67), described recently the properties of one polymorph, designated as sulfur XII. This form has properties characteristic of a modest semiconductor rather than an insulator, and reversion to octameric allotropes is slow under ambient conditions.

The above two papers suggest that a need for more sulfur research exists and that some highly significant observations can be and are being made. Studies which would improve our understanding of the burning characteristics of sulfur would be valuable in many applications. Is it possible that the different allotropes burn at different rates? How can we account for Raamsdonk's findings (68) that the addition of relatively large (up to 35%) amounts of sulfur to polystyrene foam can make it self-extinguishing? What mechanisms can be used to explain satisfactorily this unexpected property of sulfur?

Do bacteria oxidize the different sulfur polymorphs at the same or at different rates? An answer to this question could be important in the manufacture of degradable plastics or controlled release fertilizers.

Various organic and inorganic compounds alter the chemical, physical, and mechanical properties of sulfur. Many investigators have done outstanding work in relating changes in the viscosity to allotropic composition and in using additives to control the reversion of sulfur into crystalline modifications. However, our qualitative feel of what is going on is not an adequate substitute for quantitative knowledge concerning reaction rates, equilibrium concentrations, and the properties of each allotrope and the various mixtures of allotropes. Hopefully, information which can contribute significantly to the development of new uses for sulfur will become available soon.

Literature Cited

(1) Thackray, M., "Melting Point Intervals of Sulfur Allotropes," *J. Chem. Eng. Data* (1970) **15** (4), 495–97.
(2) Dale, J. M., Ludwig, A. C., "Mechanical Properties of Sulfur Allotropes," *Mater. Res. Stand.* (1965) **5**, 411–17.
(3) Tuller, W. N., Ed., "The Sulphur Data Book," pp. 73–87, McGraw-Hill, New York, 1954.
(4) Tobolsky, A. V., MacKnight, W. J., "Polymeric Sulfur and Related Polymers," pp. 1–7, 117–29, Interscience, New York (1965).
(5) Schmidt, M., Wilhelm, E., "Cyclononasulphur, S_9, A New Modification of Sulfur," *J. Chem. Soc. D* (1970) **17**, 1111–12.
(6) Barnes, M. D., "Research in Sulphur Chemistry—Current Developments and Potential Applications," *Sulphur Inst. J.* (1965) **1** (1), 2–7.
(7) Dale, J. M., Ludwig, A. C., "Reinforcement of Elemental Sulphur," *Sulphur Inst. J.* (1969) **5** (2), 2–4.
(8) Abraham, H., "Asphalts and Allied Substances," Vol. 1, p. 80; Vol. 2, pp. 178–79; Vol. 3, pp. 12, 87; 6th ed., Van Nostrand, New York, 1960.
(9) Traxler, R. N., "Asphalt, Its Compositions, Properties and Uses," pp. 90–101, Reinhold, New York (1961).
(10) Traxler, R. W., Proteau, P. R., Traxler, R. N., "Action of Microorganisms on Bituminous Materials. 1. Effect of Bacteria on Asphalt Viscosity," *Appl. Microbiol.* (1965) **13** (6), 838–41.
(11) Ariano, R., "The Action of Sulphur on Bitumens," *Strade* (1941) **9**, 119.
(12) Ellis, C., "Improved Residual Asphalts," U. S. Patent **1,020,643** (March 19, 1912).
(13) McKinney, P. V., Mayberry, M. G., Westlake, H., Jr., *Ind. Eng. Chem.* (1945) **37** (2), 177.
(14) Westlake, H. E., Jr., "The Sulfurization of Unsaturated Compounds," *Chem. Rev.* (1946) **39**, 219–39.
(15) Kronstein, M., "Process for Modifying, Solidifying, and Insolubilizing Asphalts," U. S. Patent **2,560,650** (July 17, 1951).
(16) Hughes, E. C., Hardman, H. F., "Asphalts and Waxes," *Advan. Chem. Ser.* (1951) **5**, 262–77.
(17) Farmer, E. H., "α-Methylenic Reactivity in Olefinic and Polyolefinic Systems," *Rubber Chem. Technol.* (1942) **15**, 765.
(18) Kronstein, M., "Sulphur in Elastomeric Paints," *Proc. 160th Mtg. Am. Chem. Soc.*, Chicago (September 1970) **30** (2), 238.
(19) Twiss, L., *Trans. Inst. Rubber Ind.* (1928) **3**, 386; *Chem. Abs.* (1928) **22**, 2491.

(20) Van Iterson, *Intern. Assoc. Rubber Cultiv. Netherland Indies Comm. Netherland Gov. Inst. Advis. Rubber Trade & Ind.* (1918) (Pt. 7), p. 239; *Chem. Abs.* (1919) **13**, 386.

(21) Cranker, K. R., Perrine, V. H., "A Non-Blooming High Temperature Resistant Vulcanizing Agent for Natural and Synthetic Rubbers," *Rubber Age* (1957) **81**, 113–16.

(22) Nicolau, A., Garrigues, C., Brossel, M., "Bitumen Binders Containing Sulfur for Road Surfacing," Ger. Patent **2,016,568** (Oct. 15, 1970).

(23) Welborn, J. Y., Babashak, J. F., Jr., "A New Rubberized Asphalt for Roads," *Proc. Amer. Soc. Civil Eng.* (HW2) (1958) **84**, 1651.

(24) Bacon, R. F., Bencowitz, I., "Method of Paving," U. S. Patent **2,182,837** (Dec. 12, 1939).

(25) Sommer, A., "Process of Preparing Building Compositions," U. S. Patent **2,372,230** (March 25, 1945).

(26) Hughes, E. C., Stine, H. M., "Process of Treating Asphalt to Improve its Adhesion," U. S. Patent **2,673,164** (March 23, 1954).

(27) Touro, F. J., Wiewiorowski, T. K., "Viscosity-Chain Length Relationship in Molten Sulfur Systems," *J. Phys. Chem.* (1966) **70** (1), 239–41.

(28) Quarles Van Ufford, J. J., Zwavel en Bitumen, *Thesis* Hogeschool Te Delft (Oct. 23, 1963), 80 pp.

(29) Griffin, R. L., Simpson, W. C., Miles, T. K., "The Influence of Composition of Paving Asphalt on Viscosity, Temperature, Susceptibility, and Durability," *J. Chem. Eng. Data* (1959) **4**, 349.

(30) Briggs, D. K. H., Croft, J. A., "Brittle Point of Road Tar," *J. Appl. Chem.* (1969) **19**, 12–14.

(31) Corbett, L. W., "Composition of Asphalt Based on Generic Fractionation, Using Solvent Deasphaltening, Elution-Adsorption Chromatography, and Densimetric Characterization," *Anal. Chem.* (1969) **41** (4), 576–79.

(32) Hubbard, P., Gollomb, H., *Proc. Ass. Asphalt Paving Technol.* (Dec. 6, 1937) **9**, 165.

(33) Hughes, E. C., Faris, R. B., Jr., *Proc. Ass. Asphalt Paving Technol.* (1950) **19**, 329.

(34) Lee, D. Y., Csanyi, L. H., "Hardening of Asphalt During Production of Asphaltic Concrete Mixes," *J. Mater.* (September 1968) **3** (3), 538–55.

(35) Litehiser, R. R., Schofield, H. Z., "Progress Report on Brick Road Experiments in Ohio," *Proc. 16th Ann. Mtg. Highway Res. Bd.* (November 1936), pp. 182–92.

(36) Metcalf, C. T., "Bituminous Paving Composition," Brit. Patent **1,076,866** (July 26, 1967).

(37) Speer, T. L., "Bituminous Pavement," U. S. Patent **3,239,361** (March 8, 1966).

(38) Erdmann, E., *Ann.* (1908) **362**, 133.

(39) Farmer, E. H., "Ionic and Radical Mechanisms in Olefinic Systems, with Special Reference to Processes of Double-Bond Displacement, Vulcanization and Photo-Jelling," *Trans. Faraday Soc.* (1942) **38**, 356.

(40) Gallagher, T. P., "Process for Impregnating Porous Bodies with a Solid Fusible Substance," U. S. Patent **3,341,355** (Sept. 12, 1967).

(41) Kakos, M. J., Fitzgerald, J. V., "Ceramic Tile," U. S. Patent **3,208,190** (Sept. 25, 1965).

(42) Van der Akker, J. A., Wink, W. A., "An Experimental Paper House," *Pap. Ind. Pap. World* (1948) **30** (2), 231–38.

(43) Harris, R. S., "Structural Shapes of Reinforced Sulfur and Method of Producing Same," U. S. Patent **3,183,143** (May 15, 1961).

(44) Hancock, C. K., "Plasticized Sulfur Composition for Traffic Marking," *Ind. Eng. Chem.* (1954) **46** (11), 2431–35.

(45) McKee, R. C., "Corrugated Board and Method of Making Same," U. S. Patent **2,568,349** (Sept. 18, 1951).

(46) Duecker, W. W., Schofield, H. Z., "Results from the Use of Plasticized Sulfur as a Jointing Material for Clay Products," *Amer. Ceram. Soc. Bull.* (1937) **16** (11), 435.

(47) Payne, C. R., Duecker, W. W., "Chemical Resistance of Sulphur Cements," *Trans. Amer. Inst. Chem. Eng.* (1940) **36** (1), 91–111.

(48) "Evaluation of Concrete and Related Materials for Desalination Plants," *U.S.D.I. BuReclam. Gen. Rep. No. 37B2* (Aug. 1968) 146, 152.

(49) Payne, C. R., Duecker, W. W., "Construction with Sulphur Cement," *Chem. Met. Eng.* (1940) **47** (1), 20–21.

(50) Esclamadon, C., Signouret, J. B., Labat, Y., "Plastic Sulfur Compositions," Ger. Patent **2,004,305** (August 13, 1970).

(51) Signouret, J., Audouze, B., Barge, J., "Plasticized Sulphur," U. S. Patent **3,384,609** (May 21, 1968).

(52) Williams, R., "Plasticized Sulfur Compositions," U. S. Patent **3,453,125** (July 1, 1969).

(53) Kane, J. C., "Sprayable Sulfur Road Marking Compounds," U. S. Patent **3,447,941** (June 3, 1969).

(54) Louthan, R. P., Zoeller, R. J., Hillman, D. A., "Plasticized Sulfur—A Low Cost Thermoplastic Highway Marking Material," *Speech, Am. Soc. State Hgwy. Officials Mtg.* (Autumn 1968).

(55) Dale, J. M., "Sulphur-Fibre Coatings," *Sulphur Inst. J.* (1965) **1** (1), 11–13.

(56) Johnston, V., "New Techniques," *Monthly Rev.* Federal Reserve Bank of San Francisco (June 1970), pp. 130–31.

(57) Hubbard, S. J., "Feasibility Study of Masonry Systems Utilizing Surface-Bond Materials," *U. S. Dept. Army Tech. Rep. No. 4-43* (July 1966) 20–22, 33–35.

(58) Ludwig, A. C., "Utilization of Sulphur and Sulphur Ores as Construction Materials in Guatemala," *U. N. Rep. No. TAO/GUA/4* (July 14, 1969).

(59) Testa, R. B., Anderson, G. B., "The Use of Sulphur in Housing Construction: An Exploratory Study," Columbia Univ., Dept. Civ. Eng. & Eng. Mech. (December 1969).

(60) Meyer, B. (Ed.), "Elemental Sulfur," Preface, Chap. 18, Interscience, New York (1965).

(61) Signouret, J., Barge, J., "Slowly Burning Sulphur Materials," Can. Patent **839,812** (April 21, 1970).

(62) Dale, J. M., Ludwig, A. C., "Feasibility Study for Using Sulphur-Aggregate Mixtures as Structural Material," *Sw. Res. Inst., TR No. AFWL-TR-66-57* (1966).

(63) Frusti, R. A. J., "Sulphur-Aggregate Concrete," *Mil. Eng.* (1967) **387**, 27.

(64) Crow, L. J., Bates, R. C., "Strengths of Sulfur-Basalt Concretes," *U.S.D.I., BuMines Rep. Invest. 7349* (March 1970).

(65) "Sulfur Helps Crack Problem of Cracking Concrete," *Chem. Marketing Newspaper (Oil, Paint, Drug Reptr.)* (Feb. 2, 1970) 5.

(66) Ludwig, A. C., Pena, L. H., Jr., "The Use of Sulphur to Control Reflective Cracking," *Air Force Civ. Eng.* (1969) **10** (4), 2.

(67) Vezzoli, G. C., Zeto, R. J., "Ring-Chain High-Pressure Polymorphic Transformation in Sulfur and the Accompanying Change from Insulating to Modest Semiconducting Behavior," *Inorg. Chem.* (1970) **9** (11), 2478–84.

(68) Van Raamsdonk, G. W., "Elementary Sulfur as Flame-Retardant in Plastic Foams," U. S. Patent **3,542,701** (Nov. 24, 1970).

RECEIVED March 5, 1971.

INDEX